HUMAN-COMPUTER INTERACTION SYMPOSIUM

IFIP – The International Federation for Information Processing

IFIP was founded in 1960 under the auspices of UNESCO, following the First World Computer Congress held in Paris the previous year. An umbrella organization for societies working in information processing, IFIP's aim is two-fold: to support information processing within its member countries and to encourage technology transfer to developing nations. As its mission statement clearly states,

> *IFIP's mission is to be the leading, truly international, apolitical organization which encourages and assists in the development, exploitation and application of information technology for the benefit of all people.*

IFIP is a non-profitmaking organization, run almost solely by 2500 volunteers. It operates through a number of technical committees, which organize events and publications. IFIP's events range from an international congress to local seminars, but the most important are:

- The IFIP World Computer Congress, held every second year;
- Open conferences;
- Working conferences.

The flagship event is the IFIP World Computer Congress, at which both invited and contributed papers are presented. Contributed papers are rigorously refereed and the rejection rate is high.

As with the Congress, participation in the open conferences is open to all and papers may be invited or submitted. Again, submitted papers are stringently refereed.

The working conferences are structured differently. They are usually run by a working group and attendance is small and by invitation only. Their purpose is to create an atmosphere conducive to innovation and development. Refereeing is less rigorous and papers are subjected to extensive group discussion.

Publications arising from IFIP events vary. The papers presented at the IFIP World Computer Congress and at open conferences are published as conference proceedings, while the results of the working conferences are often published as collections of selected and edited papers.

Any national society whose primary activity is in information may apply to become a full member of IFIP, although full membership is restricted to one society per country. Full members are entitled to vote at the annual General Assembly, National societies preferring a less committed involvement may apply for associate or corresponding membership. Associate members enjoy the same benefits as full members, but without voting rights. Corresponding members are not represented in IFIP bodies. Affiliated membership is open to non-national societies, and individual and honorary membership schemes are also offered.

HUMAN-COMPUTER INTERACTION SYMPOSIUM

IFIP 20th World Computer Congress, Proceedings of the 1st TC 13 Human-Computer Interaction Symposium (HCIS 2008), September 7-10, 2008, Milano, Italy

Edited by

Peter Forbrig
University of Rostock
Germany

Fabio Paternò
ISTI-CNR
Italy

Annelise Mark Pejtersen
Center of Cognitive Systems Engineering
Denmark

Editors
Peter Forbrig
University of Rostock
Germany

Fabio Paternò
ISTI-CNR, Pisa
Italy

Annelise Mark Pejtersen
Risoe National Laboratory
Roskilde, Denmark

p. cm. (IFIP International Federation for Information Processing, a Springer Series in Computer Science)

ISSN: 1571-5736 / 1861-2288 (Internet)
ISBN: 978-1-4419-3513-7 e-ISBN: 978-0-387-09678-0

Printed on acid-free paper

springer.com

IFIP 2008 World Computer Congress (WCC'08)

Message from the Chairs

Every two years, the International Federation for Information Processing hosts a major event which showcases the scientific endeavours of its over one hundred Technical Committees and Working Groups. 2008 sees the 20th World Computer Congress (WCC 2008) take place for the first time in Italy, in Milan from 7-10 September 2008, at the MIC - Milano Convention Centre. The Congress is hosted by the Italian Computer Society, AICA, under the chairmanship of Giulio Occhini.

The Congress runs as a federation of co-located conferences offered by the different IFIP bodies, under the chairmanship of the scientific chair, Judith Bishop. For this Congress, we have a larger than usual number of thirteen conferences, ranging from Theoretical Computer Science, to Open Source Systems, to Entertainment Computing. Some of these are established conferences that run each year and some represent new, breaking areas of computing. Each conference had a call for papers, an International Programme Committee of experts and a thorough peer reviewed process. The Congress received 661 papers for the thirteen conferences, and selected 375 from those representing an acceptance rate of 56% (averaged over all conferences).

An innovative feature of WCC 2008 is the setting aside of two hours each day for cross-sessions relating to the integration of business and research, featuring the use of IT in Italian industry, sport, fashion and so on. This part is organized by Ivo De Lotto. The Congress will be opened by representatives from government bodies and Societies associated with IT in Italy.

This volume is one of fourteen volumes associated with the scientific conferences and the industry sessions. Each covers a specific topic and separately or together they form a valuable record of the state of computing research in the world in 2008. Each volume was prepared for publication in the Springer IFIP Series by the conference's volume editors. The overall Chair for all the volumes published for the Congress is John Impagliazzo.

For full details on the Congress, refer to the webpage http://www.wcc2008.org.

Judith Bishop, South Africa, Co-Chair, International Program Committee
Ivo De Lotto, Italy, Co-Chair, International Program Committee
Giulio Occhini, Italy, Chair, Organizing Committee
John Impagliazzo, United States, Publications Chair

WCC 2008 Scientific Conferences

TC12	**AI**	Artificial Intelligence 2008
TC10	**BICC**	Biologically Inspired Cooperative Computing
WG 5.4	**CAI**	Computer-Aided Innovation (Topical Session)
WG 10.2	**DIPES**	Distributed and Parallel Embedded Systems
TC14	**ECS**	Entertainment Computing Symposium
TC3	**ED_L2L**	Learning to Live in the Knowledge Society
WG 9.7 **TC3**	**HCE3**	History of Computing and Education 3
TC13	**HCI**	Human Computer Interaction
TC8	**ISREP**	Information Systems Research, Education and Practice
WG 12.6	**KMIA**	Knowledge Management in Action
TC2 **WG 2.13**	**OSS**	Open Source Systems
TC11	**IFIP SEC**	Information Security Conference
TC1	**TCS**	Theoretical Computer Science

IFIP

- is the leading multinational, apolitical organization in Information and Communications Technologies and Sciences
- is recognized by United Nations and other world bodies
- represents IT Societies from 56 countries or regions, covering all 5 continents with a total membership of over half a million
- links more than 3500 scientists from Academia and Industry, organized in more than 101 Working Groups reporting to 13 Technical Committees
- sponsors 100 conferences yearly providing unparalleled coverage from theoretical informatics to the relationship between informatics and society including hardware and software technologies, and networked information systems

Details of the IFIP Technical Committees and Working Groups can be found on the website at http://www.ifip.org.

Contents

Preface

The IFIP World Computer Congress (WCC) is one of the most important conferences in the area of computer science and a number of related Human and Social Science disciplines at the worldwide level and it has a federated structure, which takes into account the rapidly growing and expanding interests in this area. Human-Computer Interaction is now a mature and still dynamically evolving part of this area, which is represented in IFIP by the Technical Committee 13 on HCI. We are convinced that in this edition of WCC, which takes place for the first time in Italy, it will be interesting and useful to have a Symposium on Human-Computer Interaction in order to present and discuss a number of contributions in this field.

There has been increasing awareness among designers of interactive systems of the importance of designing for usability, but we are still far from having products that are really usable, and usability can mean different things depending on the application domain. We are all aware that too many users of current technology feel often frustrated because computer systems are not compatible with their abilities and needs with existing work practices. As designers of tomorrow technology, we have the responsibility of creating computer artefacts that would permit better user experience with the various computing devices, so that users may enjoy more satisfying experiences with information and communications technologies. This has raised new research areas, such as ambient intelligence, natural interaction, end user development, work analysis and cultural and social interaction.

The interest to the conference has been positive in terms of submissions and participation for an event at its first edition. We have received 40 contributions, and 12 have been accepted as long papers and 8 as short ones. The selection has been carried out carefully by the International Programme Committee. The result is a set of interesting and stimulating papers that address such important issues as Elicitation and Evaluation, Collaboration and Visualization, E-learning, Task Models and Interaction The interest shown in the conference has truly been world-wide: if we consider both full and short papers we have authors from West and East Europe, South America, Japan, and South Africa. The final programme of the symposium includes one technical invited speaker: Luca Chittaro from University of Udine on Interacting with Visual Interfaces on Mobile Devices. This is a topic, which is acquiring an increasing interest, given that recent years have seen the introduction of many types of computers and devices (e.g. cellphones, PDA's, etc.) and the availability of such a wide range of devices has become a fundamental challenge for designers of interactive software systems.

In general, the continuous development of new research topics in the human-computer interaction area shows how the field is able to dynamically evolve and address both new and old challenges. All the results obtained are never an arrival point but they are the basis for new research and results and we hope that the IFIP Symposium on Human-Computer Interaction can contribute to this process.

April 2008,

Peter Forbrig, Fabio Paternò, Annelise Mark Pejtersen

HCIS 2008 Conference Organization

The 1st IFIP Human-Computer Interaction Symposium (HCIS 2008)

is a co-located conference organized under the auspices of the
IFIP World Computer Congress (WCC) 2008
in Milano, Italy

Peter Forbrig
HCIS 2008 Co-Chair
Peter.forbrig@uni-rostock.de

Fabio Paternò
HCIS 2008 Co-Chair
fabio.paterno@isti.cnr.it

Annelise Mark Pejtersen
HCIS 2008 Co-Chair
ampcse@mail.dk

HCIS 2008 Program Committee

Julio Abascal, Spain
Nikos Avouris, Greece
Simone Diniz Junqueira Barbosa, Brazil
Anke Dittmar, Germany
Luca Chittaro, Italy
Torkil Clemmensen, Denmark
Maria Francesca Costabile, Italy
Michael Harrison, United Kingdom
Philippe Palanque, France
Phil Gray, United Kingdom
Paula Kotze, South Africa
Effie Law, Switzerland
Gitte Lindgaard, Canada
Monique Noirhomme, Belgium
Lars Oestreicher, Sweden
Gerd Szwillus, Germany
Gerrit van der Veer, The Netherlands
Chengqing Zong, China

Interacting with Visual Interfaces on Mobile Devices

Invited Talk

Luca Chittaro

HCI Lab, Dept. of Math and Computer Science, University of Udine, Italy
chittaro@dimi.uniud.it, http://hcilab.uniud.it

Abstract: This invited talk will discuss some major issues in developing visual interfaces for mobile devices as well as demonstrate a number of new mobile visual applications and interaction techniques that we have developed in domains as diverse as tourism, health & fitness, navigation, emergency response, geospatial and architectural visualization. It will also deal with the issue of providing designers with new tools to study the behavior of users of these new mobile applications.

Keywords: human-computer interaction, mobile devices, visual interfaces, information visualization, user studies

1. Contents of the talk

Mobile devices such as mobile phones and PDAs are increasingly used to support information needs of users on the move. As a consequence, information spaces that have been traditionally available only to desktop and laptop users (e.g., documents, pictures, web pages, maps, large databases,…) are moving to small screens as well, presenting application designers with new challenges. Indeed, the technical limitations of mobile devices combined with the peculiar needs of users on-the-go require a careful design of applications that are specifically thought for mobile devices and users [1]. As an example, the common form factors of mobile devices constrain screen space to a small fraction of what is available on a desktop. A typical 240x320 pixels display of a PDA has less than 1/16 the area of a typical 1280x1024 desktop display (see Fig. 1). Such size limitation makes it extremely difficult for users to navigate information spaces that do not fit a single screen, unless appropriate techniques to simplify interaction and navigation are provided.

Please use the following format when citing this chapter:

Chittaro, L., 2008, in IFIP International Federation for Information Processing, Volume 272; *Human-Computer Interaction Symposium*; Peter Forbrig, Fabio Paternò, Annelise Mark Pejtersen; (Boston: Springer), pp. 1–5.

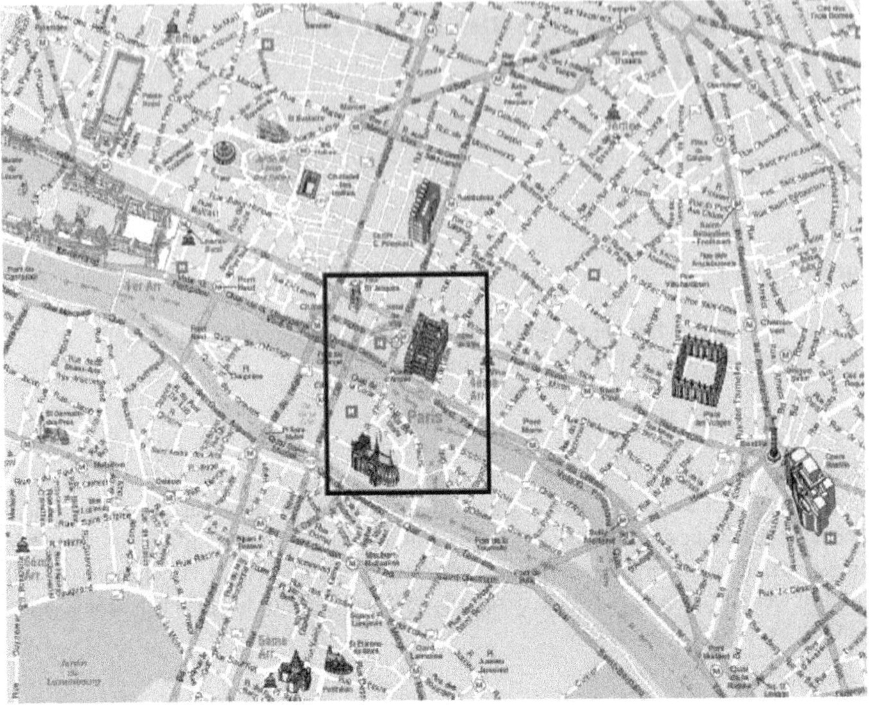

Figure 1 Comparison between the size of a typical 240x320 PDA screen (the area highlighted by the black rectangle) and a common 1280x1024 desktop screen (the whole picture).

The recent availability of mobile devices with increasingly powerful graphics capabilities is making it possible to develop novel visual interfaces, based on interactive 2D (or even 3D) graphics, to help users on the move in dealing more quickly and easily with larger amounts of information. Limited cognitive resources and safety of mobile users are an additional motivation to employ mobile graphics effectively as a way to provide information at-a-glance that is easily understood with less cognitive resources and distracts the user as less as possible from her surrounding environment.

Mobile visual interfaces become even more innovative and provide functionalities that were unavailable on desktop systems when they are integrated with various sensors (e.g., GPS, accelerometers, heart rate monitors, pulseoximeters,...) that allow one to adapt the behavior of the application according to position in space (location-awareness) and other parameters (context-awareness). In this way, the mobile device becomes able to choose what to show and how to show it on the display based on what is happening to the user as well as the physical world that surrounds her. For example, we developed a mobile personal trainer (see Fig. 2), called MOPET [2], that monitors user's position and physiological parameters in outdoor sports activities to present functionalities such as location-aware maps augmented with visualizations of users' performance or context-aware fitness advice and 3D demonstrations of exercises.

Figure 2 A wrist-worn context-aware system supports a user in outdoor fitness activities [2]

This invited talk will discuss some major issues in developing visual interfaces for mobile devices as well as demonstrate a number of new mobile applications and interaction techniques that we have developed in domains as diverse as tourism[3], health & fitness [2], navigation [4], emergency evacuation [5], geospatial and architectural visualization [6]. After an initial discussion of the peculiarities of the mobile context that motivate research, we will analyze some of the issues involved in designing a mobile visual interface, especially focusing on the so-called presentation problem [1,7,8,9]. The talk will illustrate a number of interaction and visualization solutions we have recently proposed and evaluated on users, such as the Zoom-Enhanced Navigator (ZEN) to explore information spaces on small screens [4], the MAGDA system to analyze datasets of geo-referenced elements in the field [3], and specific techniques for managing the off-screen objects [10] and the icon cluttering [11] problems. Then, location-awareness and context-awareness aspects will be introduced, considering very different application domains such as fitness training [2] or emergency response [5], and applications based on different kinds of sensors.

Finally, the talk will deal with the issue of providing designers with new tools to study the behavior of users of these new mobile applications. In particular, it will illustrate three different tools we have recently proposed for that purpose: the MOBREX tool [12] to log user interface actions on the mobile device and then visually analyze them on a desktop system, the VU-Flow [13] and the MOPET Analyzer [14] tools for the visual analysis of users' data collected through sensors connected to mobile devices.

References

1. Chittaro L. Visualizing Information on Mobile Devices, IEEE Computer, Vol. 39, No. 3, pp. 34-39 (2006).
2. Buttussi F., Chittaro L., MOPET: A Context-Aware and User-Adaptive Wearable System for Fitness Training. Artificial Intelligence In Medicine Journal, Vol. 42, No. 2, pp. 153-163 (2008).
3. Burigat S., Chittaro L., Interactive Visual Analysis of Geographic Data on Mobile Devices based on Dynamic Queries, Journal of Visual Languages and Computing, Vol. 19, No. 1, pp. 99-122 (2008).
4. Burigat S., Chittaro L., Gabrielli S., Navigation Techniques for Small-screen Devices: an Evaluation on Maps and Web pages, International Journal of Human-Computer Studies, Vol. 66, No. 2, pp. 78-97 (2008).
5. Chittaro L., Nadalutti. D. Presenting Evacuation Instructions on Mobile Devices by means of Location-Aware 3D Virtual Environments, Proceedings of MOBILE HCI 2008: 10th International Conference on Human-Computer Interaction with Mobile Devices and Services, ACM Press, New York, in press (2008).
6. Mulloni A., Nadalutti D., Chittaro L., Interactive Walkthrough of Large 3D Models of Buildings on Mobile Devices, Proceedings of Web3D 2007: 12th International Conference on 3D Web Technology, ACM Press, New York, pp. 17-25 (2007).

7. Gustafson, S., Baudisch, P., Gutwin, C, and Irani, P. Wedge: Clutter-Free Visualization of Off-Screen Locations, Proceedings of the CHI 2008 Conference on Human factors in computing systems, ACM Press, New York, pp. 787-796 (2008).
8. Bederson B.B., Clamage A., Czerwinski M.P., Robertson G.G. DateLens: A fisheye calendar interface for PDAs, ACM Transactions on Computer-Human Interaction, vol. 11, no.1, pp. 90-119 (2004).
9. Ware C. Information Visualization: Perception for Design, 2nd Edition, Morgan Kaufmann, San Mateo, CA, (2004).
10. Burigat S., Chittaro L., Gabrielli S., Visualizing Locations of Off-Screen Objects on Mobile Devices: A Comparative Evaluation of Three Approaches, Proceedings of MOBILE HCI 2006: 8th International Conference on Human-Computer Interaction with Mobile Devices and Services, ACM Press, New York, pp. 239–246 (2006).
11. Burigat S., Chittaro L., Decluttering of Icons based on Aggregation in Mobile Maps, In Meng L., Zipf A., Winter S. (eds), Map-based Mobile Services - Design, Interaction and Usability, Springer, Berlin, pp. 13-32 (2008).
12. Burigat S., Chittaro L., Ieronutti L., Mobrex: Visualizing Users' Mobile Browsing Behaviors, IEEE Computer Graphics and Applications, Vol. 28, No. 1, pp. 24-32 (2008).
13. Chittaro L., Ranon R., Ieronutti L., VU-Flow: A Visualization Tool for Analyzing Navigation in Virtual Environments, IEEE Transactions on Visualization and Computer Graphics, Special Issue on Visual Analytics, Vol. 12, No. 6, pp. 1475-1485 (2006).
14. Nadalutti, D., Chittaro, L. Visual Analysis of Users' Performance Data in Fitness Activities, Computers & Graphics, Special Issue on Visual Analytics, Vol. 31, No. 3, pp. 429-439 (2007).

Identification Criteria in Task Modeling

Josefina Guerrero García, Jean Vanderdonckt, and Christophe Lemaigre

Belgian Laboratory of Computer-Human Interaction (BCHI)
Louvain School of Management (LSM), Université catholique de Louvain (UCL)
Place des Doyens, 1 – B-1348 Louvain-la-Neuve (Belgium)
E-mail: josefina.guerrero@student.uclouvain.be, {jean.vanderdonckt, christophe.lemaigre}@uclouvain.be

Abstract: Task modeling consists of a fundamental activity that initiates user-centered design in user interface development. It is therefore important to reach the best task model possible and that the task modeling activity remains consistent when the task modeler changes. For this purpose, this paper introduces a set of criteria in order to identify tasks during task modeling in an unambiguous way that results into a task model exhibiting desired properties of quality such as completeness, consistency. In addition, starting and stopping criteria provide designers with guidance on when and how to start and finish the task modeling.

Keywords: Process, task identification criteria, task modeling.

1. Introduction and Related Work

Task modeling (Duursma, 1993; Paterno, 1999) is probably one of the most central activities to conduct in order to ensure user-centered design of interactive systems. A task model is supposed to capture most elements describing how a task is carried out by a particular user in a given context of use or in a given scenario (Limbourg, 2003).

In general, the purpose of model-based design, for instance (Calvary, 2003; Limbourg, 2003; Santoro, 2002; Sinnig, 2007; Traetteberg, 1999; Vanderdonckt, 2003), of User Interfaces (UIs) is to identify high-level models which allow designers to capture specifications of interactive applications from a more abstract level than the implementation level at which the future application will be developed. This allows designers to concentrate on important design aspects without being influenced by the implementation constraints. In particular, the task model's goal is to capture specifications of how a task is carried out in a given context of use (Calvary, 2003) (i.e., a triple user-computing platform-physical environment).

Please use the following format when citing this chapter:

García, J.G., Vanderdonckt, J. and Lemaigre, C., 2008, in IFIP International Federation for Information Processing, Volume 272; *Human-Computer Interaction Symposium*; Peter Forbrig, Fabio Paternò, Annelise Mark Pejtersen; (Boston: Springer), pp. 7–20.

A *task* is an activity that should be performed in order to reach a goal. A *goal* (in UI model-based design) is a desired modification of state or an inquiry to obtain information on the current state of an interactive application (Paterno, 1999). One of the advantages of task modeling lies in its characterization of the logical activities that an interactive application must support independently of any underlying technology or implementation. Task modeling has become today a widely recognized activity in the UI development life cycle. Several task models are precisely defined and are adequately made editable through software (Limbourg, 2003): Hierarchical Task Analysis (HTA) is supported by Architect, Méthode Analytique de Description de tâche (MAD) is supported by SUIDT (Aït Ameur, 2003); Goals, Operators, Methods, and Selectors (GOMS) is supported by QGOMS (Beard, 1996), Task Knowledge Structures (TKS) is supported by ADEPT, Groupware Task Analysis (GTA) is supported by EUTERPE (van der Veer, 2000), ConcurTaskTree (CTT) (Paterno, 1999) is supported by CTTE-editor (Santoro, 2002), Diane is supported by the Diane+ editor, ISOLDE is supported by an eponym editor. Despite these recent advances, task modeling still remains a challenging problem for the following reasons:

- Although there is more or less a consensus about its definition, about the information to be captured in a task model, and about the tool usage, yet there exists a significant gap on the means to be used to obtain such a task model.
- In the literature, there is little or no methodological guidance on how to obtain such a task model. When some guidance is provided, it mainly consists of syntactical rules to get a task model without any defect. These rules are completely independent of the domain of human activity. Formal validation of a task model is considered very important (Aït Ameur, 2003), but its external validation (i.e., with respect to the users' needs) is equally important.
- This often results in many variations in the task model obtained in the end: different people may produce different, possibly inconsistent, task models for the same design problem because they do not share the same perception or rules; a same person (e.g., a task analyst, a task modeler) may produce task models with different levels of details depending on the design problem; even more, a same person can produce different task models for the same design problem over time.
- People experience some trouble in identifying the points where to start the task modeling and where to stop it. Until when should we proceed with task modeling such as decomposition and refinement?
- People may diverge on their interpretation of what needs to be captured in a task model and what not, in particular what makes a task and what does not make a task? Several different interpretations of what a task model is and what task modeling should be co-exist without reaching any consensus (Limbourg, 2003).
- This problem is even more acute when task modeling is conducted in the context of a larger design problem such as workflow modeling (Guerrero, 2008):

it is difficult to distinguish what is workflow specific from what is task specific.

- The relationship between a task model and use cases that depict a particular scenario as a task model instance is obvious, but yet hard to obtain (Constantine, 1999), although some method exists that establishes this type of relationship (Santoro, 2002).
- Equally important are the design rationale techniques used to argue and to reason about the task modeling decisions (Lacaze, 2006; MacLean, 1991).

Since there is no apparent need to conduct any research for the model part (the task model has gained today a precise and shared definition) or any development for the tool support (excellent software are publicly available for this purpose, such as CTTE-editor (Paterno, 1999)), we believe that there is still some research to be conducted for improving the methodological guidance for conducting task modeling. This guidance should be of course independent of any task model definition or tool since the task modeling activity should be achieved consistently by any person.

In order to fit this purpose, namely by addressing the aforementioned shortcomings, this paper will in Section 2: define the underlying models, define an expanded task life cycle, and define a set of criteria for identifying a task with respect to other concepts. Section 3 will exemplify a case study based on this methodological guidance and how software that has been developed for this purpose may facilitate applying and structuring this guidance. Section 4 will provide a qualitative cost-benefit analysis of this guidance and present some future avenues of this research.

2. Toward Methodological Guidance for Task Modeling

The next sub-sections will define, respectively, the three dimensions that are typically found in a software development method (Bodart, 1989): the underlying models, the development life cycle, and the method part. These concepts will be exemplified in Section 3.

2.1 Underlying Models

A single model cannot capture all aspects relevant to task modeling. If everything would have been concentrated in such a model, then it would not adhere to the *Principle of Separation of Concerns*. It is therefore necessary to have a family of interconnected models so as to preserve correlativity between models. It is not needed to have all models involved in any development life cycle, but it is desirable to have such a family of models in order to capture all desired aspects.

The Cameleon Reference framework structures the UI development life cycle into four subsequent layers (Calvary, 2003): task and domain models, abstract user interface (AUI) model, concrete user interface (CUI) model, and final user interface (FUI). In order to properly capture various aspects in respective models, this framework has been expanded (Fig. 1) in order to incorporate the concepts of workflow, process, and resource (Guerrero, 2008). A workflow is decomposed into processes, which are in turn decomposed into tasks. Each task could be supported thanks to a resource model, in which three types of resources can be found: human resources (i.e. a user stereotype), material resources (e.g., hardware, network, machines), and immaterial resources (e.g., software, operating system). More details about the attributes and methods of these classes could be found in (Guerrero, 2008). Fig. 1 only represents the UML class diagram of this metamodel without any attributes or methods. This expanded framework will serve as a reference framework in order to identify concepts that are relevant to these different models, if needed. If a scenario or a use case does not incorporate any element relevant to a particular model, then this model is simply not created.

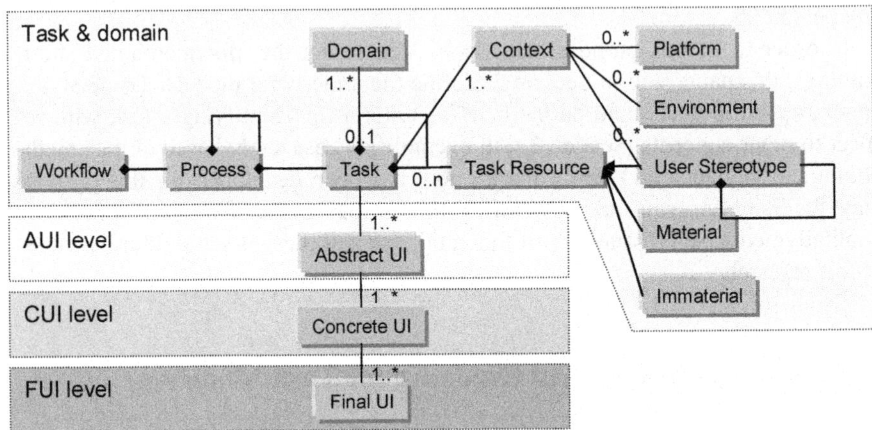

Figure 1. The four levels of the Cameleon Reference framework expanded in this paper expanded from (Calvary, 2003).

This expanded framework allows us to support a total integration of models in the context of model-based UI development: vertical integration is obtained when different models at different levels of abstraction are interconnected and; horizontal integration when different models describing different aspects at the same level of abstraction are interconnected. In our case, horizontal integration is obtained at the first level of Figure 1 since all models are clearly separated and interconnected. Vertical integration is ensured since the cornerstone model, i.e., the task model, initiates the rest of the subsequent levels of abstraction (i.e., AUI, CUI, and FUI). Now that the relationship between a task model and other related models have been outlined, we would like to define the central concept of task in such a way that its identification becomes as precisely as possible. Therefore, we define:

- A *task model* describes the task analyst's view of the end users' interactive tasks while interacting with the system, where a task is any operation unit that is carried out in the same time-space frame with the same set of resources for the same information set.
- A *process model* describes how tasks are arranged in time, space, and resources so as to form basic more elaborate operation units satisfying transitive disclosure (i.e., having a clear entry point, a middle portion, and an exit point). Each process consisting of a number of tasks and a set of conditions that determine the execution order of the tasks, and task relationships.
- A *workflow model* describes the flow of the work inside, outside, and between organizations. In other words, a workflow model is aimed at representing the flow of work inside and outside organizational units in terms of tasks that describe the way humans perform tasks to accomplish a goal.

From the above definitions, we will first deduce an expanded task life cycle in the next sub-section and precise criteria in order to identify what a task is, what a process is, and what a workflow is, depending on the presence of these concepts in a case study, a design problem, or a scenario of interest.

2.2 Expanded Task Life Cycle

Tasks are dynamic entities whose life-cycle can be described with a small quantity of significant states. Changes in those states, along with transitions between those states, are produced by the stimulus. On the one hand, there are some particular states of interest that result from the definition: cancellation, suspend/resume, among others. On the other hand, we have to separate the definition of a task from the means to allocate it to resources and from the states where the task is indeed executed. Each task can therefore benefit from the following actions thus resulting in the life cycle of Figure 2:

- *Cancel*: any task can be cancelled, once started at any moment.
- *Delegated*: any task can be delegated to another resource (e.g., another user stereotype) once it has been allocated or initiated.
- *Finished*: any task is said to be finished when the goal is reached.
- *Undo*: any task can be undone once initiated.
- *Redo*: any task that has been undone can be redone.
- *In course*: any task could be executed.
- *Repeat*: any task that has been accomplished can be repeated as many times as necessary.
- *Review*: any completed task can be reviewed before it is considered finished.

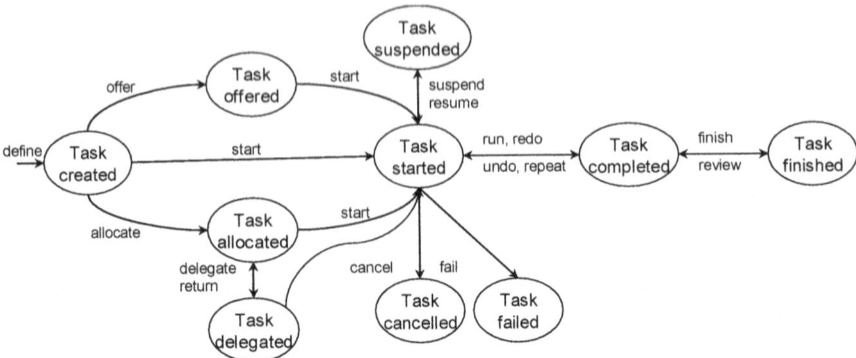

Figure 2. Expanded task life cycle.

2.3 Identification Criteria for a Task, a Process, and a Workflow

Since a task is defined as an operation executed while four dimensions remain constant (i.e., time, space, resources, information), any variation of any of these four dimensions, taken alone or combined, thus generate a potential identification of a new task in the task modeling activity. A task is like "a play", which is decomposed into acts, which are in turn decomposed into scenes (Figure 3a). In a piece, an act is a unit during which time, space, and actors remain constant, even across scenes. Variation to any of these dimensions means that another act has begun. Similarly for a task, any variation of time, space, set of resources or information set will mean a change of task. It is important that identifying a new task is independent of any task decomposition. In the IDA (Interactive Design Approach) method (Bodart, 1989), the decomposition is limited in that it is only possible to decompose into four levels: a project is recursively decomposed into interactive applications, which are in turn decomposed into phases (with one level only), which are in turn decomposed into functions (Figure 3b). In our method, we do want to stay independent of any decomposition (in order to accommodate any task modeling approach). Therefore, we do not limit the number of decomposition levels, as represented by the recursive aggregations in Figure 1 for processes and tasks.

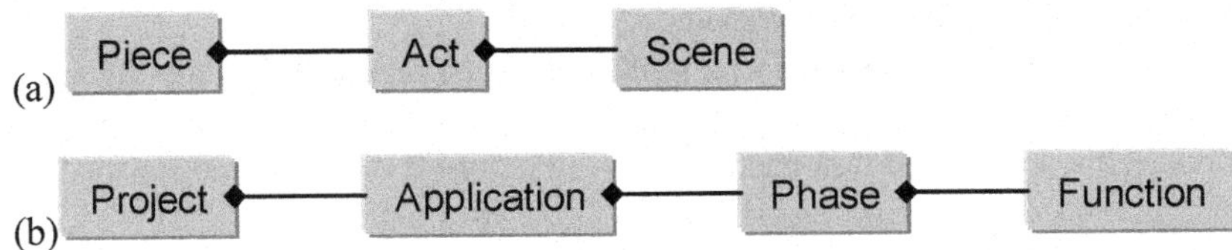

Figure 3. Some decomposition: (a) a play; (b) a project into sub-units.

From the definition of a task, we deduce the following identification criteria:

1. *Change of space* (or change of location): when the scenario indicates a change of location of the operations, a change of task may occur. Therefore, any scenario fragment like "in the headquarters, the worker does ..., then in the local agency, the worker does..." indicated a change of space, therefore a change of task. The main location where the task is carried out takes the advantage. In case of collaborative or cooperative tasks, the main location is considered to detect whether there is any change of location.
2. *Change of resource*: when the scenario suggests that new or different resources are exploited, a change of task may occur. We distinguish three categories of resources from Fig. 1:
   a. *Change of resource of type "User stereotype"*: when another user stereotype appears in the scenario may indicate that there is a change of task. For example, "a clerk does ..., then an employee files the results of ...". The two different names for two different users indicate a change of task. This reasoning always forces identifying who is the main responsible person for carrying out a task. Again, in case of collaborate or cooperative tasks, the main user stereotype involved in the task is selected.
   b. *Change of resource of type "material"*: when another material resource appears in the scenario, a change of task may occur. For example, "a clerk enters the customer's data on a PocketPC, and then takes a picture with a mobile phone camera" indicates two tasks resulting from the usage of two different resources, here a PocketPC and a mobile phone. This should not be confused with a task that is performed on different computing platforms, like in the context of a multi-device UIs (Santoro, 2002). Thus, any significant change of software and/or hardware resource may indicate that there is a change of task.
   c. *Change of resource of type "immaterial"*: when another immaterial resource appears in the scenario, a change of task may occur. For example, "a network administrator uses a specific software to check network status; s/he uses another software to update the computers of the network". The two different types of software involved indicate a change of task.
3. *Change of time*: when the scenario indicates a different time period in which the task is performed, a change of task may occur. We differentiate four criteria resulting from any potential change of time:
   a. *Existence of an interruption*: when the task is interrupted by an event that changes the time period. For instance, "an employee registers every incoming complaint. After registration a form is sent to the customer who returns the form within two weeks". A task can be interrupted for many reasons, such as an error, an external event, a dynamic task.
   b. *Existence of a waiting point*: when in the development of a task there is a moment where it is necessary to wait that something occurs for continuing, a change of task may occur. We have two types of a waiting points:
      i. *Waiting point of type "decision"*: when a determination arrives at after consideration of a question, a change of task may occur. For example, "after the preparation of a flight plan, the pilot will take the decision to fly". A decision can be made by a human agent, a system

agent or in a mixed-initiative way. In any case, there could be as many different tasks as there are different alternatives coming out the decision. In this way a decision could result into two tasks after the decision or multiple tasks (like in a "case of").

  ii. *Waiting point of type "accumulation"*: when there is necessary to create a waiting list for some information, a change of task may occur. For instance, "due to a car accident, more complaints arrived yesterday at the insurance agency and the employee had to register all incoming complaints to send as a group to directors". The accumulation may be related to documents (or messages, or data) or to processes (e.g., a repetition of tasks).

c. *Permanence of execution unit*: when the task execution depends of the results of at least two previous asynchronous tasks. For instance, "the results of an insurance complaint are delivered to the client when the complaint manager provides whether the complaint applies or not and when the evaluator provides the estimated cost".

d. *Periodicity of execution*: when there are different periodicities established to execute tasks, then a change of task may occur. For example, "every Monday the employee does a backup of the information". This criteria is often detected when one can determine that the frequency of one task is different from the frequency of another. For instance, a backup (automatic) task could be incremental every day and full each Friday. In this case, we separate two tasks (incremental vs total backup) because their frequencies are different: every day vs. every Friday.

4. *Change of nature*: when the scenario represents a change of category, a change of task may occur. A task may have any of the following nature: manual, automated, interactive or mechanical. Any change of this nature may indicate a change of task. For instance, "first a secretary types a letter in the computer (interactive), after a printer prints the text (automatic) and finally the manager signs the letter (manual)".

When doing task modeling, it is important to decide how far the decomposition of tasks is to proceed. This depends of course of the context and purpose of task modeling; however some stopping criteria based on the task life cycle (Fig. 2) are:

1. For *horizontal stopping*: when the task is finished, or the task is canceled or the task failed.
2. For *vertical stopping*: when a task can be performed in a simple and well-determined way (i.e. the task cannot be decomposed in sub-tasks), when the task is executed by a software system and we do not intend to replace this system with anything else (Duursma, 1993).

Table 1 lists a set of parameters to identify a workflow and a process following the methodology applied in the task identification. Notice that our method assumes that a textual description of the problem has been gathered by any method, for instance, interviews, and the author assumes this information as complete. The way this information is collected and its completeness in not in the scope of this paper, so, as to compare our method with other task analysis method. The above guidelines covers the challenges described in Section 1.

Table 1. Identification criteria

	Time	Space (location)	Resource	Type
Workflow	Series of time periods	Different locations; same organization	Same or different groups of resources	-
Process	Series of time periods	Different locations	Within groups, group as a whole, or among groups	Primary (production), secondary (support), or tertiary (managerial)
Task	Same time period	Same location	One or two types of resources	User, interactive, system, abstract, or machine task

3. Feasibility study

The literature demonstrated that there is a rich set of task modeling approaches and practices, for instance (Limbourg, 2003; Paterno, 1999; Rosson, 2007; Sinnig, 2007; Traetteberg, 1999; van der Veer, 2000). The set of identification criteria that was defined in the previous section is not intended to replace any of these approaches and practices. Rather, the identification criteria are interpreted as a complementary tool that disambiguates the tasks involved in task modeling. These criteria do not change the task analysis activity per se, but, again, attempts to facilitate the identification of tasks.

In order to demonstrate that this approach based on identification criteria is feasible, we have conducted a series of case studies, some of which being reported in (FlowiXML, 2008). Empirical evidence of method's worth should be one of the first requirements for its acceptance. For space reasons we present a simplified version of the case study related to an insurance company.

3.1 Case Study

An insurance company offers insurances of different types as: medical, life, and accidents. Each type of insurance is processed in different section of the company, in different way, and by different resources. To process claims that result from car accidents, the company uses the following procedure:

- Every claim is registered by a desk clerk. He introduces the customer's full name, policy number, and the problem's details in the insurance company system.
- After the registration, an employee checks the insurance's information as the status, the insurance's coverage, and the damage history.
- Also, he phones to the garage to get the cost of the damages.
- After executing these tasks a decision is made by the manager, with two possible outcomes, positive or negative.
- If the decision is positive, then the insurance company will pay. An accountant handles the payment.
- The secretary of the insurance company sends a letter to the customer. Also, a letter will be sent to the customer when the decision is negative explaining the causes of the rejection.

From the text, we identify that each type of insurance is a process; all of them form a workflow. In the particular case of the accident insurance process, it is possible to classify principal tasks, add an identifier and a name. We can specify if the task needs that a previous one is finished, make a brief description of the task, list the identification criteria and identify the nature of the task. The previous steps to classify tasks are supported by a task classifier, software tool specialized (Fig. 4).

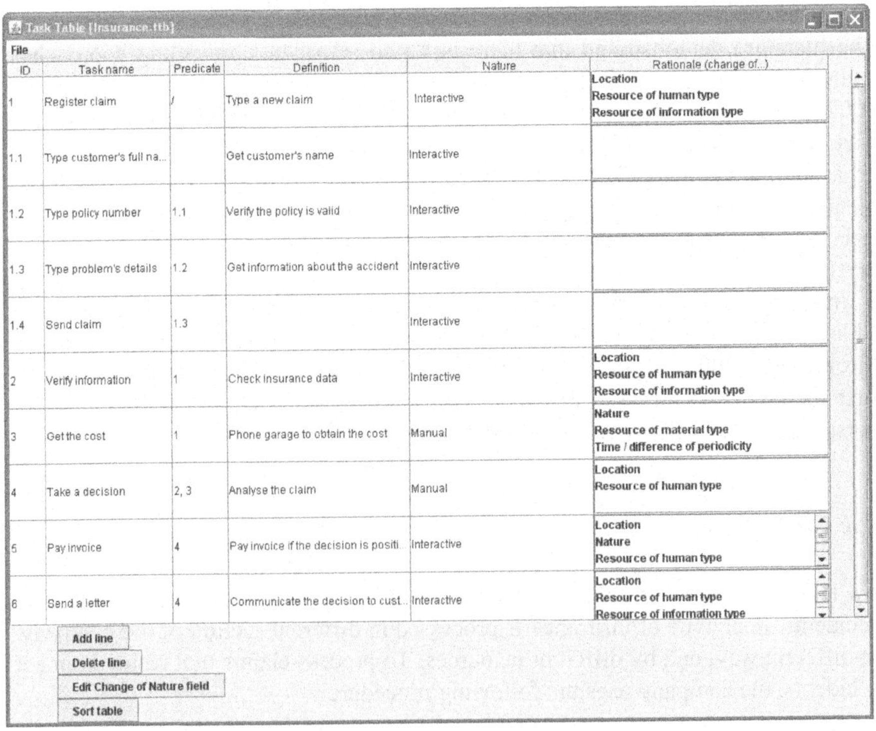

Task Table [Insurance.ttb]

File

ID	Task name	Predicate	Definition	Nature	Rationale (change of...)
1	Register claim	/	Type a new claim	Interactive	Location Resource of human type Resource of information type
1.1	Type customer's full na...		Get customer's name	Interactive	
1.2	Type policy number	1.1	Verify the policy is valid	Interactive	
1.3	Type problem's details	1.2	Get information about the accident	Interactive	
1.4	Send claim	1.3		Interactive	
2	Verify information	1	Check insurance data	Interactive	Location Resource of human type Resource of information type
3	Get the cost	1	Phone garage to obtain the cost	Manual	Nature Resource of material type Time / difference of periodicity
4	Take a decision	2, 3	Analyse the claim	Manual	Location Resource of human type
5	Pay invoice	4	Pay invoice if the decision is positi...	Interactive	Location Nature Resource of human type
6	Send a letter	4	Communicate the decision to cust...	Interactive	Location Resource of human type Resource of information type

Add line

Delete line

Edit Change of Nature field

Sort table

Figure 4. Tool to classifier tasks.

As we observe, a task can be decomposed in sub-tasks, for instance Task 1 *Register claim* is decomposed in sub-tasks 1.1 *Type customer's full name*, 1.2 *Type policy number*, 1.3 *Type problem's details*, and 1.4 *Send claim*. Even though task 2 and task 3 are executed by the same resource, they present a change of resource type (immaterial and material), a change of nature (interactive and manual), and a change of time. In order to execute task 4, task 2 and task 3 should be finished; also there is a change of space and resource. If the result of task 4 is positive, then task 5 could be executed with a change of space, resource of type user stereotype, immaterial, and exist a waiting point of type decision. Task 6 will be executed after task 4, it is necessary a waiting point of type decision, a change of space, and a change of resource.

After the identification and classification of tasks, processes and workflow, we can represent all these components in a workflow editor tool (Figure 5). The workflow describes as the work in organization flows by defining models of: process (what to do?) and tasks (how to do it?). For each process a task model can be specified to describe in detail how the task is performed.

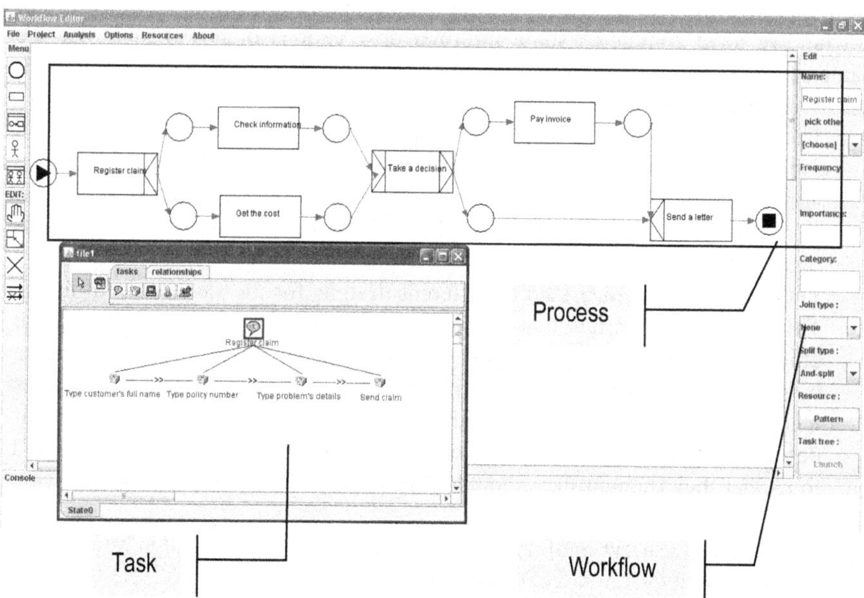

Figure 5. Workflow model.

3.2 Validation

To validate this approach, we are working on the theoretical and empirical considerations in order to address the internal and external validations respectively. The empirical validation is being conducted with multiple case studies research available on www.usixml.org and in (Vanderdonckt, 2005)

- **Bank credit for a car**. In this work we will focus just in analyzing the workflow of how to ask a credit to buy a car. We consider this service interesting as it involves at least three different organizations (the bank, car agencies, car buyer-seller company), related in a strategic join venture alliance. Car agencies and the car seller-buyer agency are benefit from the bank credits.
- **Organization of a Triathlon**. The organization of this event is wide and full of flows of information. In short, people wanting to organize an event have to contact different companies and fulfill the needs of the athletes and the spectators.
- **Order personalized compression stockings over Internet**. The case study is situated in the phlebology domain. It deals wit an Internet order system, allowing the ordering of personalized support stockings. The main idea of t is system is to calculate a 3D model of the customer's legs from a series of digital pictures taken from his/her legs. This model will be sent, coupled with a specific order, via Internet to the manufacturing department.

So far, we have conducted some informal observations of how people use the same set of criteria for supporting task modeling, but these observations only indicate some properties:

- The interpretation of the identification criteria is usually perceived as straightforward by people who have used it because the criteria have been defined very distinctively from each other ; in addition, the fact that a task should, in principle, keep constant the space, the time, and the resources is an easy-to-remember and fast-to-apply procedure.
- The set of criteria may lead to different designs for the same task to model. This does not mean that that the resulting designs are inconsistent with each other, but simply that the designer has chosen a particular design alternative by prioritizing the identification criteria. We observed however that designers who are using the same criteria with the same priorities tend to reach a similar design. This should be made clear to designers because some of them reported to us that they thought they should reach a single design if they use the same set of criteria.
- The usage of criteria reinforces the need for recording the design rationale as recommended in (Lacaze, 2006; MacLean, 1991).
- The usage of identification criteria also permits to separate a task from its sub-tasks. If one decides that this action should become a high-level task, then the corresponding criteria should be used for this purpose; when one decides that this action should become a low-level task (e.g., a function, a leaf node in the task model, or a sub-task), then other criteria are exploited to justify this design decision.

4. Conclusion

In this paper, we have introduced a set of precise criteria that can be used in order to identify a task in a textual scenario and to distinguish a task from other concepts like process and workflow which are located at another level in the hierarchy, but at the same level of abstraction. The main advantage of these criteria is that they can be used for any task modeling activity, whatever the task model notation or method used. The second main advantage is that a convergence across designers can be observed when the same textual scenario is given to different persons, thus increasing the internal consistency of the resulting task model. The same advantage is propagated to other models at the same level of abstraction. For this purpose, the Cameleon Reference framework (Calvary, 2003) has been expanded in order to illustrate vertical and horizontal integrations.

In order to support this activity in scenario-based design, a piece of software has been implemented in Java 1.5 that enables designers to conduct the modeling approach that is compatible with this expanded framework and by applying the set of identification criteria in a systematic way. Each time a task has been properly identified, i.e., with at least one identification criteria (multiple criteria could be used to identify the same task), it is then subject to deep task modeling, in connection with the other aspects such as process and workflow. The software then automatically generates a report that can be later used to justify aspects of task modeling within a process, or process modeling within a workflow.

Work in progress includes the evaluation of task analysis methods compare with the one presented in this paper. Dealing with user errors or problems during interaction will be examined to determine their impact in the method. Finally, further empirical evaluation will be conducted confronting two groups of task analysts, the first group, designing a solution without using the guidelines and the second using them. Thus, these results will contribute to the credibility of the proposal.

Acknowledgments. We gratefully acknowledge the support of the SIMILAR network of excellence (http://www.similar.cc), the European research task force creating human-machine interfaces similar to human-human communication of the European Sixth Framework Programme (FP6-2002-IST1-507609) and the CONACYT program (www.conacyt.mx) supported by the Mexican government. We also greatly thank the anonymous reviewers for their constructive feedback that was helpful for improving this manuscript.

References

Aït Ameur, Y., Baron, M., and Girard, P.: Formal Validation of HCI User Tasks. In: Proc. of the Int. Conf. on Software Engineering Research and Practice SERP'2003 (Las Vegas, June 23-26, 2003), pp. 732-738, CSREA Press (2003).

Beard, D., Smith, D., and Danelsbeck, K.: QGOMS: A direct-manipulation tool for simple GOMS models. In: Proc. of ACM Conf. on Human factors in Computing Systems CHI'96 (Vancouver, April 14-18, 1996), pp. 25-26, ACM Press, New York (1996).

Bodart, F., and Pigneur, Y.: Conception assistée des systèmes d'information : modèles, méthode, outils, Dunod, Paris (1989).

Calvary, G., Coutaz, J., Thevenin, D., Limbourg, Q., Bouillon, L., and Vanderdonckt, J.: A Unifying Reference Framework for Multi-Target User Interfaces, Interacting with Computers, 15(3), pp. 289-308 (June 2003).

Constantine, L.L., an Lockwood, L.A.D.: Use cases in task modeling and user interface design. In: Proc. of ACM Conf. on Human Factors in Computing Systems CHI'99 (Pittsburgh, May 15-20, 1999), p. 352, ACM Press, New York (1999).

Duursma, C.: Task Model definition and Task Analysis process, ESPRIT Project P5248 KADS-II KADS-II/M5/VUB/RR/004/1.1c, Vrije Universiteit Brussel, Brussels (1993).

Guerrero, J., Vanderdonckt, J.: FlowiXML: a Step towards Designing Workflow Management Systems, Journal of Web Engineering, 4(2), pp. 163-182 (2008).

Lacaze, X., Palanque, P., Barboni, E., Bastide, R., and Navarre, D.: From Dream to Realitiy: Specificities of Interactive Systems Development with respect to Rationale Management. In: A.H. Dutoit, R. McCall, I. Mistrik, B. Paech (Eds.), Rationale Management in Software Engineering, pp. 155-172, Springer, Heidelberg (2006).

Limbourg, Q., and Vanderdonckt, J.: Comparing Task Models for User Interface Design. In: D. Diaper, N. Stanton, N. (Eds.), The Handbook of Task Analysis for Human-Computer Interaction, pp. 135-154, Lawrence Erlbaum Associates, Mahwah (2003).

MacLean, A., Young, R.M., Bellotti, V., and Moran, T.: Questions, Options and Criteria: elements of design space analysis, Journal on Human Computer Interaction, 6(3-4), pp. 201-250 (1991).

Paterno, F., and Mancini, C.: Developing task models from informal scenarios. In: Proc. of ACM Conf. on Human Aspects in Computing Systems CHI'99 (Pittsburgh, May 15-20, 1999), ACM Press, New York (1999).

Rosson, M.B., and Carroll, J. M.: Scenario-based Design. In: Sears, A., Jacko, J.A. (Eds.), The human-computer interaction handbook: fundamentals, evolving technologies, and emerging applications, CRC Press (2007).

Santoro, C., Mori, G., and Paterno, F.: Ctte: Support for developing and analyzing task models for interactive system design, IEEE Transactions on Software Engineering, 28(9), pp. 797-813 (September 2002).

Sinnig, D., Chalin, P., and Khendek, F.: Towards a Common Semantic Foundation for Use Cases and Task Models. In: Proc. of the 1st Int. Workshop on Formal Methods for Interactive Systems FMIS'2006, Electronic Notes in Theoretical Computer Science, Vol. 183, pp. 73-88 (11 July 2007).

Trætteberg, H.: Modelling Work: Workflow and Task Modelling. In: Proc. of 3rd Int. Conf. on Computer-Aided Design of User Interfaces CADUI'1999 (Louvain-la-Neuve, October 21-23, 1999), pp. 275-280, Kluwer Academics Publishers, Dordrecht (1999).

Vanderdonckt, J., Furtado, E., Furtado, V., Limbourg, Q., Silva, W., Rodrigues, D., and Taddeo, L.: Multi-model and Multi-level Development of User Interfaces. In: A. Seffah, H. Javahery, (Eds.), Multiple User Interfaces - Cross-Platform Applications and Context-Aware Interfaces, pp. 193-216, John Wiley, New York (November 2003).

Vanderdonckt, J.: A MDA-Compliant Environment for Developing User Interfaces of Information Systems. In: Proc. of 17th Conf. on Advanced Information Systems Engineering CAiSE'05 (Porto, June 13-17, 2005), Lecture Notes in Computer Science, Vol. 3520, pp. 16-31, Springer, Heidelberg (2005).

van der Veer, G., van Welie, M.: Task based groupware design: putting theory into practice. In: Proc. of the ACM Conf. on Designing Interactive Systems: Processes, Practices, Methods, Techniques DIS'2000 (New York, August 17-19, 2000), pp. 326–337, ACM Press, New York (2000).

Website FlowiXML. Available via UsiXML. http://www.usixml.org/index.php?mod=pages&id=40. Accessed April 14th, 2008.

HCI-Task Models and Smart Environments

Maik Wurdel*, Stefan Propp*, and Peter Forbrig

University of Rostock, Department of Computer Science
Albert-Einstein-Str. 21, 18059 Rostock, Germany
{maik.wurdel, stefan.propp, peter.forbrig}@uni-rostock.de

Abstract The paper discusses the idea of using HCI-task models to support smart environments. It introduces a collaborative task modeling language CTML that allows the specification of collaboration and comprehensive dependencies in an OCL-like style. Additionally some ideas are presented that allow informing users and usability experts about the state of actors within smart environments. The paper provides the first results of a prototypical implementation.

Keywords. HCI, Task Models, Model-based Usability Evaluation

1. Introduction

In the domain of HCI task analysis and modeling is a mature research area. Task models are used to elicit requirements in early stages of development by describing how people achieve goals by performing a set of tasks. However, in recent years, task models have also been employed for system design. Exemplary in the research field of model-based user interface (UI) development task models serve as initial model for model-based processes. In contrast in the research field of smart environments HCI task models have only been used barely. From our point of view this fact is quite surprising because smart environments comprise a vast complexity in terms of task performance of users. A thorough understanding of the tasks users are executing within such environments is a precondition to deliver an appropriate assistance.

In this paper we focus on using task models in smart environments to, first, understand the envisioned assistance and, second, to track the task performance during runtime. This approach consists of two major components: (1) the collaborative task modeling language to model the behavior of actors within smart environments and (2) usability evaluation methods to provide usability experts with evaluation support and to inform actors about the current state of the system.

*Supported by a grant of the German National Research Foundation (DFG), Graduate School 1424, Multimodal Smart Appliance Ensembles for Mobile Applications (MuSAMA)

Please use the following format when citing this chapter:

Wurdel, M., Propp, S. and Forbrig, P., 2008, in IFIP International Federation for Information Processing, Volume 272; *Human-Computer Interaction Symposium*; Peter Forbrig, Fabio Paternò, Annelise Mark Pejtersen; (Boston: Springer), pp. 21–32.

Additionally we introduce our tool support which allows for modeling and simulation of collaborative tasks and their execution environment. Our simulation environment allows for interactively walk through the designed artifact, while conducting the usability evaluation.

The remainder of the paper is structured as follows: in Secion 2 we stress some background information to tasks and smart environments. Section 3 introduces our Collaborative Task Modelling Language (CTML) and the corresponding tool support which is followed by Section 4 where usability evaluation methods for smart environments are discussed. Finally we draw the conclusion and give an outlook for future research avenues.

2. Modeling Tasks in Smart Environments

Within smart environments tasks are barely carried out in isolation, but have to be synchronized with other users' tasks. Some tasks cannot be started while others are still in progress. To motivate our research we illustrate the challenges of smart environments concering task models by a scenario. Afterward we reiterate through existing approaches of task modeling and its employements.

The session chair Dr. Smith introduces herself and defines the topic of the session. Afterwards she gives the floor to the first speaker who sets up her equipment, the laptop switches to presentation mode and the speaker starts with the talk. During the presentation the audience accesses additional information related to the talk using their personal devices. While the meeting proceeds the personal devices provide guidance and offer related information according to the current talk and the meeting progress. The chairman interrupts the speaker since she overruns her time slot. The plenum is asked for some brief questions which are answered by the speaker. Eventually the chairman closes the talk and announces the next one. Subsequent talks are given in a simliar same manner.

We consider a smart environment as location where people are collaborating using a set of stationary and mobile devices. The devices are supposed to support the users' tasks which have to be performed to achieve a certain goal (like giving a talk). Addtionally interaction with the environment is performed in a much broader way (Shirehjini, 2007) then in desktop applications. The initiative can be expressed explicitly or implicitly. An implicit interaction is understood as an action not performed to interact with the environment but interpreted by the system. Ideally an implicit proactive meeting assistant for instance does not wait for an explicit user command, but senses movements and gestures of the user via sensors to derive the assumed user intention and automatically provides support for the expected next task.

Based on the introductionary scenario we can elicit the key characeteristic of smart environments from the view of task modeling: (1) A vast amount of potential tasks supported by a dynamic set of devices. (2) The temporal order of tasks depends on the collaboration of actors within the environment. (3) The state of the smart environment (defined as composed state based on each device) can furthermore restrict or enable the execution of a certain set of tasks.

Diverse notations for task models have been introduced (GOMS, HTA, CTT, WTM (Bomsdorf, 2007; van Welie, 1998)). Even though they differ in terms of presentation, expressiveness, level of formality and granularity they all share the same following basic principle: tasks are arranged hierarchically representing the decomposition of tasks and tasks are performed to achieve a certain goal. The decomposition of tasks stops when an atomic level is reached: the action. It builds the fundamental execution unit.

The most common notation ConcurTaskTrees (CTT) supports, amongst others, the concept of temporal relations which restricts the valid sequences of tasks to achieve a certain goal. Another asset of this notation is its tool support: CTTE (Mori, 2002). Various extensions have been introduced: Examplarily in (Bomsdorf, 2007; Klug, 2005) an action is not seen as atomic anymore, but defined by a life cycle. This defines a task more precisely which is employed to trigger events. The first approach does not consider a temporal operator as state chart whereas the latter does not consider abortion or skipping of tasks.

Modeling cooperation of users in terms of task models has been addressed by CCTT (Collaborative ConcurTaskTrees) (Mori, 2002). Similar to the corporative task modeling language presented in this paper, CCTT uses a role-based approach. A CCTT specification consists of multiple task trees. One task tree for each involved user role and another as a "coordinator" that specifies the collaboration and global interaction between involved user roles.

Model-based usability evaluation approaches, like RemUSINE (Paterno, 2007), capture interaction events to derive the performed user interaction on an abstract task-based level. A trace of task events contains qualitative information about accomplished tasks, as well as quantitative measures about durations of fulfilled tasks. Analysis approaches comprise e.g. (Malý, 2007; Paterno, 2007). The suggested visualizations are based on a linear time-based scale. However, the visualization approach presented in this paper applies a semantic lens to focus on a certain period of time.

After reviewing existing approaches we introduce our specification language which comprises the characteristics of smart environments for task modeling.

3. CTML – the Collaborative Task Modeling Language

CTML is based on the idea that in limited and well-defined domains the behavior of an actor can be approximated through her role and, second, the behavior of each role can be adequately expressed by an associated collaborative task expression.

According to this statement we correspondingly define a collaborative task model as a tuple consisting of a *set of actors*, a *set of roles*, a *set of devices* and a *set of collaborative task expressions* (one for each role) where each actor belongs to one or more role(s).

Definition 1: (Collaborative Task Model). A collaborative task model G is a tuple $G=\langle A,R,T,D,F,a,r,p\rangle$ where:

A,R,T,D are non empty sets of actors, roles and collaborative task expressions and devices. F is the set of features of the model consisting of elements of the following kind: $\langle key, value\rangle$

$a{:}A \rightarrow P(R)$ is a function that associates an actor with a set of roles.

$r{:}R \rightarrow T$ is a bijective function that associates a role with a task model.

$p{:}A \cup D \rightarrow F$ is a relation associating features to the actors and devices

Each collaborative task expression has the form of a task tree, where nodes are either tasks or temporal operators. Each task is attributed with a (unique) identifier, a precondition and an effect. Intuitively, the precondition defines a required state of the collaborative environment for executing the task, whereas an effect denotes the resulting state after having executed the task. Addtionally temporal operators restrict the potential sequences of task performance.

Definition 2: (Collaborative Task Expression). A collaborative task expression CTE is a tuple $CTE=\langle T,h\rangle$ where,

T is a non-empty set of tasks of the form $\langle id, precondition, effect\rangle$

$h{:}\ T \rightarrow List(T) \times Op$, with $Op = \{[], |=|, |||, |>, [>, >>, *, \#, opt\}$ is a function that maps a task t to an ordered list of tasks and a temporal operator.

The former represents the children of task t, whereas the latter denotes the execution order of the children according to the given definition in (Sinnig, 2007).

We say a collaborative task expression is *well formed* if the corresponding task tree is connected and free of cycles such that each task (except for the root task) has exactly one parent. Moreover we demand that if a task has more than two children it is associated with an n-ary operator. If a task has exactly two or one child(ren) it is associated with a binary or unary operator respectively. Leaf tasks are not related to a temporal operator by the function h. This definition results in a bi-parit graph whose vertexes are either of T or Op. The function h defines the edges of the graph.

In Fig. 1 a subset of the collaborative task model for the introductory example is given that was interactively created using the CTML Editor. For the sake of readability for each task only the hierarchical breakdown and temporal relations are shown in a CTT-like style. Our editor is able to present temporal relations as nodes or in the CTT-style. Preconditions and effects have been omitted for the example below. An overview of the entire specification is given in the lower left corner.

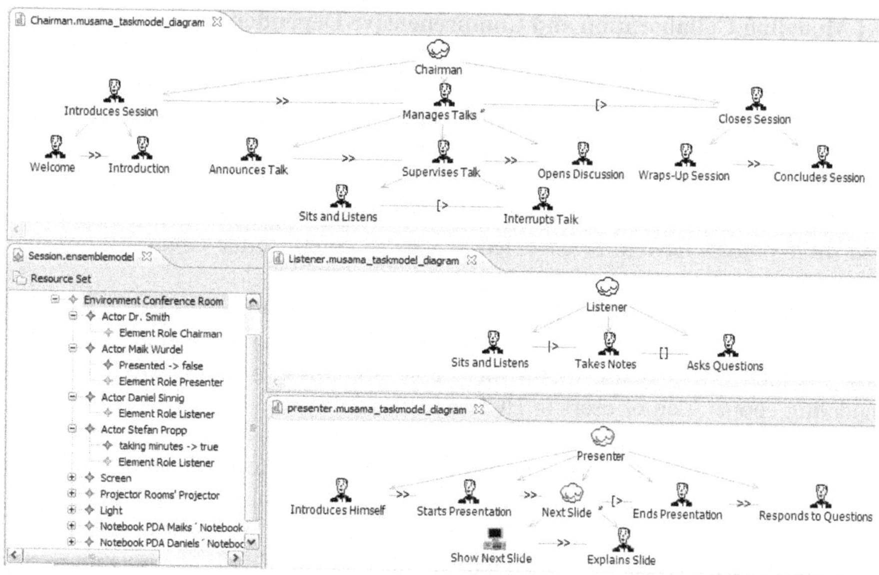

Fig. 1. Specification of the Example Using the CTML Editor

At runtime, for each active actor, an instance of the corresponding collaborative task expression (identified by the assigned role) is created. To implement the operational semantics of a CTML model all instance task expressions are translated to sets of communicating state charts. On the one hand task state charts are responsible for checking the precondition and manipulating the system to achieve its effect. On the other hand temporal operator state charts implement the semantics of its corresponding operator. In particular they mediate messages between parent and child task.

As already mentioned above a collaborative task expression can be interpreted as a bi-parit graph whose nodes are either tasks or temporal operators. Since each task and operator is mapped to a corresponding state chart a bi-parit graph of state charts is created. Edges can be seen as communication channels (similar to CSP (Hoare, 1978)). Thus state charts only communicate with adjacent state charts. This approach supports the concept of separation of concerns and helps to reduce complexity of the communication. Since each collaborative task expression is transformed into a set of communicating state charts and a colabortative task model is defined by a number of task expressions we can accordingly say that the runtime model is defined by a network of sets of communicating state charts.

Up to now we have explained the syntax and rationale of our model. In the next section we elaborate on the modeling of cooperation and dependencies of the environment according to our formally defined model.

3.1 Modeling Collaboration and Comprehensive Dependencies

From the very beginning of classical task modeling the HCI community has regarded objects of the domain as highly related to the task performance. Artifacts and tools are often mandatory to accomplish a task successfully. This fact also applies to smart environments, even though they can be physical objects, devices (stationary or mobile) as well as digital information. Dependencies between tasks and these objects have to be modeled as well to comprise the complexity of the scenario.

Moreover collaboration between actors within the environments has to be supported as well. As pointed out in the scenario there exist various interrelation of task between different users even in simple scenarios. Existing approaches of modeling cooperation of users in HCI lacks flexibility and linkage to objects and devices dependencies. CTML supports both requirements by using an OCL-like language to specify additional execution constraints and effects of tasks.

To execute a task the logical statement in the precondition has to hold which can be either based on the state of the system or the state of an actors' task execution. The abstract syntax of a precondition defined in an EBNF-like notation is as follows:

precondition = *attributePrec* | *taskPrec;*
attributePrec = *identifier DOT check;*
taskPrec = *identifier DOT task DOT state;*
identifier = *(ROLE DOT quantifier)* | *NAME;*

Note that we spare non terminals defined by char sequences (e.g. ROLE; NAME). The first (*attributePrec*) checks whether a set of properties (syntactically defined by the set *F*) has a certain value. Thus, preconditions allow for expressing dependencies of tasks and devices and/or actors. The latter (*taskPrec*) is able to express that a set of arbitrary tasks of actors are in a certain state (E.g. task t1 of actor a1 has to be started before task t2 of actor a2 is able to be started). Note that the life cycle of a task is defined in terms of a state chart whose states can be referenced in preconditions. Additionally preconditions support quantification of actors by means of roles. The meanings of the quantifier are described in Table 1.

Table 1. Semantics of Quantifiers used in Preconditions and Effects

Quantifiers		
All-Quantifier	*allInstances*	All actors of the role have to satisfy the constraint.
Exist- Quantifier	*oneInstance*	At least one actor of the role has to satisfy the constraint.
Non-Quantifier	*noInstance*	The statement holds if no actor of the role satisfies the constraint.

To illustrate the rationale of preconditions some examples for the following precondition are given in Table 2:

(1.) A presenter is allowed to start her presentation after the chairman has announced the talk.

(2.) The listeners are allowed to ask questions after the Dr. Smith has opened the discussion session.

(3.) The chairman can wrap-up the session after all presenters have finished their talk (specified by the property "presented").

Table 2. Examples of Precondition using the Different Features of the Language

#	*Role*	*Task*	*Precondition*
(1.)	*Presenter*	*StartsPresentation*	*Chairman.oneInstance.AnnouncesTalk.completed*
(2.)	*Listener*	*AsksQuestion*	*DrSmith.OpensDiscussion.completed*
(3.)	*Chairman*	*Wraps-UpSession*	*Presenter.allInstances.presented == true*

By the usage of precondition we are able to add execution constraints based on elements of the environment. This comprises the extra complexity of the domain. However to model the dynamics of such a scenario in an adequate manner the effect of a task execution has to be taken into account as well. In contrast to preconditions, effects do not check whether a logical statement holds, but specify the system state after execution the task. Similarly effects either address properties of elements or tasks of actors. The abstract syntax is illustrated here:

effect = attributeEffect | taskEffect;

attributeEffect = identifier DOT assignment;

taskEffect = identifier DOT task DOT message;

identifier = (ROLE DOT quantifier) | NAME

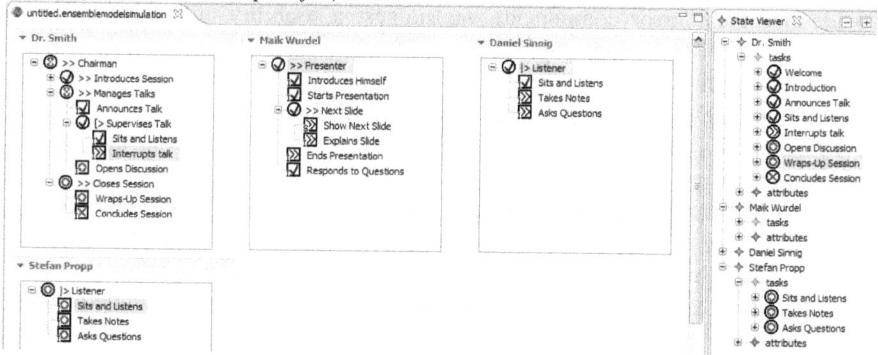

Fig. 2. CTML Simulation incorporating Preconditions and Effects

Please note that an effect on a task is not necessarily taking place since if a message is sent it is interpreted according to the potential transitions defined in the task state chart. For example it is not possible to move from state disabled to running. In this case the running message will be ignored. This avoids inconsistencies and supports the implementation of the message concept. For reasons of brevity we spare examples for effects. The simulation taking into account preconditions and effects to support collaborative task modeling is depicted in Fig. 2.

4. Usability Problems in Smart Environments

Smart environments differ from desktop applications in various aspects, which lead to an according adaptation of usability evaluation methods. Based on the charcteristics of smart environments (Section 1) we derive appropriate evaluation methods. Afterwards we show how to apply these methods for both: providing the users with information about the current state of the system and providing the usability expert with evaluation support.

4.1 Introduction to Usability Evaluation in Smart Environments

The advanced features of smart environments are able to provide a comfortable usage experience, but also introduce new possible usability issues. The reason for usability problems of proactive systems can be decomposed into four potential error components: (1) imprecise sensor values (e.g. wrong location values), (2) misinterpretations of sensor values (e.g. when applying a faulty user movement model to clean the raw sensor data), (3) intention recognition errors (e.g. when predicting the wrong user task) and (4) planning errors (e.g. when delivering the wrong functionality).

To identify these error components we suggest a usability evaluation process comprising three subsequent stages:

(1) Comparing interaction traces (Hilbert, 2000) with a predefined expected behavior to identify possible usability issues.

(2) Analysis of captured sensor data and manual annotations to investigate the reason for the problem.

(3) Investigation of the analysis metrics and visualizations to solve the issue.

Within smart environments a task can be accomplished cooperatively by a number of users by support of their different devices. In addition a certain user can start a task on one device (e.g. a mobile phone with speech input) completing the task later with another device (e.g. a laptop with keyboard). In this case separate interaction traces of the devices can hardly be compared. Therefore we suggest interpreting the interaction trace according to an underlying task model as task trace (Hilbert, 2000). A task trace is understood as arbitrary sequence of performed tasks. Deviations according to the defined temporal order of tasks may occur and need further investigation during evaluation.

Designing a usability test case comprises two activities. First the environment has to be modeled as CTML model and afterwards a usability expert defines the test plan, as it is common practice in usability evaluation.

For the execution of a usability test case we distinguish usability evaluation at different development stages. In early phases, like design, the environment is simulated as an animation of the defined CTML model (see Fig. 2). An interactive walk through helps to expose weaknesses within the designed artifacts, to revise the underlying CTML models.

After setting up the physical environment, the link to the underlying task models has to be kept to allow evaluation. We provide HTTP access to connect the smart environment and the simulation engine (Fig. 2). During a simulation every leaf task of the simulated task models can be triggered by the events “start” and “stop”. These events are internally propagated between adjacent nodes in the task model and cause the task nodes to change the state. All internal and external events are captured to build a task event trace, which is defined as a sequence of events. Each event comprises the corresponding usability test case, the task model, the task, the fired event and a success value. The captured task event trace is used to provide support for both: the usability expert for evaluation (Section 4.2) and the actors within the environment for guidance (Section 4.3). Our approach provides evaluation simultaneously to the test as well as afterwards.

4.2 Visualization and Analysis for the Usability Expert

After capturing a trace of executed tasks and the corresponding sensor data, our approach provides support for identification and analysis of usability issues. To cope with the vast amount of captured data we distinguish between two solutions: on the one hand removing data, which is out of evaluation scope, through filtering and on the other hand keeping all data, but setting focus on data of evaluation interest through aggregation.

For aggregation of the task trace we apply a semantic lens method. Analog to an optical lens a semantic lens is defined by a focus point, a size of the lens and a lens function (Griethe, 2005). Applied to a task trace, the task of interest is focussed, the size of the lens is the number of previous and successive tasks which are covered by the lens and the lens function defines how the aggregation works. The lens function defines the level of aggregation for each position within the lens area (Propp, 2007b). An example for the application of a semantic lens is shown below.

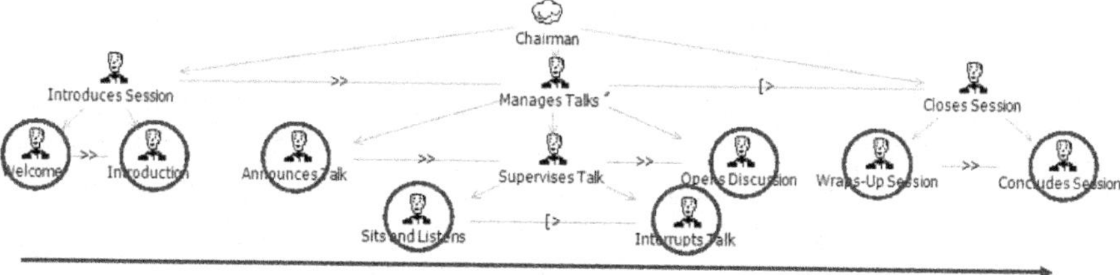

Fig. 3. Complete Example Task Trace

To continue the running example, an interaction trace for *Dr. Smith*, the chairman, is depicted in Fig. 3. The already performed tasks are highlighted. An application of a semantic lens leads to a less detailed trace in Fig. 4. In particular we set the focus on the task in the center to provide a more concise overview. The more distant the tasks are on the time scale in comparison to the focus, the higher the level of aggregation.

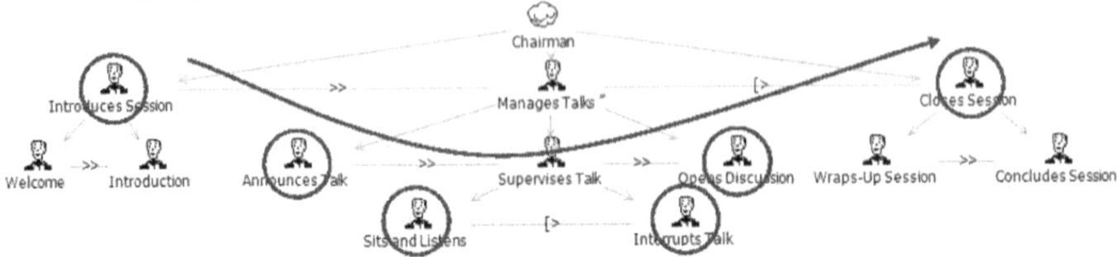

Fig. 4. Aggregated Example Task Trace

The data is captured as a trace with a time stamp for each completed task. The aggregation mechanism analyses the trace to find subsequences which have a common parent within the task tree. Depending on the focus function certain tasks are aggregated and represented by a parent or even more abstract task. The usability expert is able to choose the focus in the time scale and vary the size of focus accordingly. Adjusting the focus function provides a more or less detailed view. The filtered and aggragted task trace can be visualized with different techniques. One simple trace is depicted in Fig. 5.

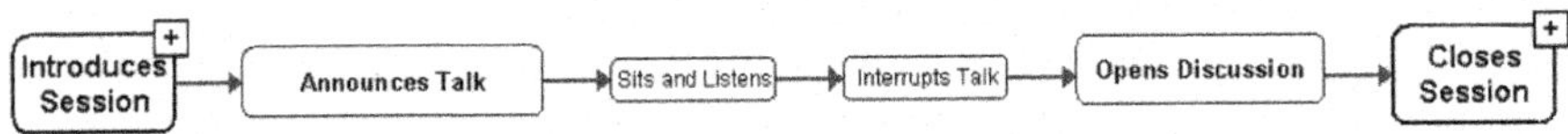

Fig. 5. Visualization of the Task Trace for a Usability Expert

Our intention is to provide specific visualization techniques for different purposes of evaluation and to have a tool box containing adaptable visualizations. Additonally we provide a timeline view to compare different users according to duration of accomplishing different tasks.

4.3 Visualization for the End User

We intend to support users with a guidance mechanism to visualize the current progress of task execution. Especially in a smart environment it is necessary to provide an overview of the current state of the system. Users might be astonished about some reactions like switching off the light. Hence it might be needed to rollback the system to prior state or forward to a new state.

Therfore we reuse the task trace to provide a history and an outlook of task execution (Propp, 2007a). Traces are prepared in the same way as for the usability expert in Section 4.2. First a filtering stage reduces accomplished tasks, e.g.

- fulfilled within the currently proceeding activity (branch of task model),
- within a predefined time interval in the past,

- with the devices that a user controls.

The subsequent aggregation step applies a semantic lens to provide a more concise overview. Additionally to the performed trace the potential future tasks are derived from the enabled task set, which contains all executable tasks at a certain moment in time. To accomplish the goal the user has a set of possibilities defined in the task model. Therefore the reasonable alternatives are already known and can be visualized as future avenues. We continue the example of the chapter 4.2 and visualize the data for user guindance in Fig. 6.

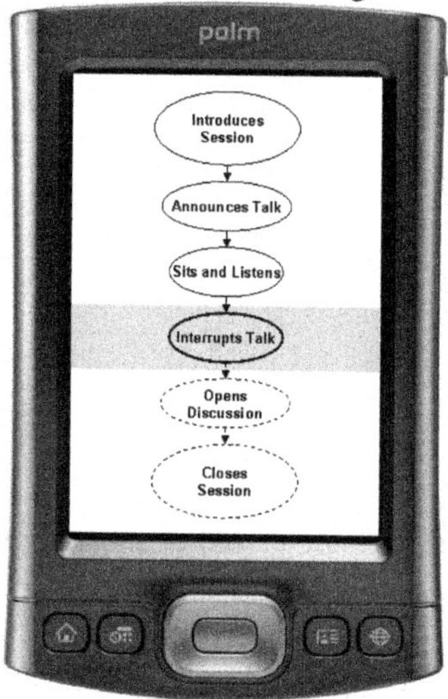

Fig. 6. Visualization of the Task Trace providing User Guidance

The example in Fig. 6 shows the chairmans' PDA to summarize the current situation of the smart environment. The task trace proceeds from top to bottom on a timeline. The focus is automatically set at the currently proceeding task, which is highlighted. The bigger shapes depict tasks at a higher level of abstraction, which are derived within the aggregation stage. The potential future tasks are visualized as dashed oval shapes. Changes within the environment are recognized by sensors and delivered to the usability framework to update the visualization accordingly.

5. Conclusion & Future Work

In this work we presented a collaborative task modeling language which can be used to model the behavior of actors in smart environments. Therefore we extended classical task modeling notation to comprise the raised complexity of the domain such as dynamic collaboration of actors and dependencies of user tasks and the environment. To enable software designers using the notation we developed an editor as well as a simulator for CTML. In the second part of this paper we elaborated on usability problems within the domain of smart environments. First we highlighted the challenges of usability in this particular domain followed by an approach which makes use of CTML as runtime engine to track the task performance of users. This approach tries to guide the user by visualizing the recent, current and potential future task during task performance. Additionally different visualization techniques are proposed which can help to evaluate task performance by usability experts.

Future research avenues comprise the evaluation of the specification at runtime by a "Wizard of Oz" experiment. This will help us to expose strengths and weaknesses of our approaches based on real data which applies for modeling as well as usability evaluation. Based on these results extending CTML will be another issue of investigation to integrate other elements of the environment. Further aspects of the usability evaluation process will be directly intregated into the modeling environment

6. References

Bomsdorf, B. (2007). "The WebTaskModel Approach to Web Process Modelling." TaMoDia **4849**: 240-253.

Griethe, H., G. Fuchs and H. Schumann (2005). A Classification Scheme for Lens Technique. WSCG (Short Papers) 2005, Plzen, Czech Republic.

Hilbert, D. M. and D. F. Redmiles (2000). "Extracting usability information from user interface events." ACM Comput. Surv. **32**(4): 384-421.

Hoare, C. A. R. (1978). "Communicating sequential processes." Commun. ACM **21**(8): 666-677.

Klug, T. and J. Kangasharju (2005). Executable Task Models. TaMoDia. Gdansk, Poland.

Malý, I. and P. Slavík (2007). Towards Visual Analysis of Usability Test Logs Using Task Models. Task Models and Diagrams for Users Interface Design**:** 24-38.

Mori, G., F. Paternò; and C. Santoro (2002). "CTTE: Support for Developing and Analyzing Task Models for Interactive System Design." IEEE Trans. Softw. Eng. **28**(8): 797-813.

Paterno, F., A. Russino and C. Santoro (2007). Remote Evaluation of Mobile Applications. TaMoDia 2007. Toulouse, France.

Propp, S. and G. Buchholz (2007a). A User Control Mechanism for Smart Appliance Ensembles. KI 2007 Workshop. Osnabrück, Germany.

Propp, S. and G. Buchholz (2007b). Visualization of Task Traces. Interact 2007 Workshop on New Methods in User-Centered System Design. Rio de Janeiro, Brasil.

Shirehjini, A. A. N. (2007). A Multidimensional Classification Model for the Interaction in Reactive Media Rooms. Human-Computer Interaction. HCI Intelligent Multimodal Interaction Environments**:** 431-439.

Sinnig, D., M. Wurdel, P. Forbrig, P. Chalin and F. Khendek (2007). Practical Extensions for Task Models. TaMoDia, Springer. **4849:** 42-55.

van Welie, M., G. van der Veer and A. Eliëns (1998). An Ontology for Task World Models. DSV-IS 98. Abingdon, United Kingdom, Springer.

Themes in Human Work Interaction Design

Rikke Orngreen[1], Annelise Mark Pejtersen[2], Torkil Clemmensen[3]

[1] Center for Applied ICT, CBS, Denmark, orngreen@cbs.dk
[2] Chair of IFIP TC 13, Denmark, ampcse@mail.dk
[3] Department of Informatics, CBS, Denmark, tc.inf@cbs.dk

Abstract. This paper raises themes that are seen as some of the challenges facing the emerging practice and research field of Human Work Interaction Design. The paper has its offset in the discussions and writings that have been dominant within the IFIP Working Group on Human Work Interaction Design (name HWID) through the last two and half years since the commencement of this Working Group. The paper thus provides an introduction to the theory and empirical evidence that lie behind the combination of empirical work studies and interaction design. It also recommends key topics for future research in Human Work Interaction Design.

Keywords: work analysis, interaction design, HCI tools, new ways of working

1. Introduction – Scope and Research Area

Technology is changing human life and work contexts in numerous ways: interfaces between collaborating individuals in advanced ICT networks, small and large-scale distributed systems, multimedia and embedded technologies, mobile technologies, and advanced "intelligent" robots. With this change towards new ways of working, an intensive demand has taken place for techniques and technologies that address contemporary issues related to communication, collaboration, learning, problem solving and information seeking in large information spaces of great variability. To address this comprehensive problem, an IFIP working group on Human Work Interaction Design (HWID) was established. Its expressed purpose was to reach a better understanding of the new challenges entailed in the design of technological support for modern, dynamic and complex work environments through a discussion of the interrelation between Work Analysis and Interaction Design within the field of Human Computer Interaction .

The main problem addressed is how we can understand, conceptualize and design for the complex and emergent contexts in which human life and work are now embroiled. This issue calls for cross disciplinary, empirical and theoretical approaches that focus on Human – Work Interaction design.

Please use the following format when citing this chapter:

Orngreen, R., Pejtersen, A.M. and Clemmensen, T., 2008, in IFIP International Federation for Information Processing, Volume 272; *Human-Computer Interaction Symposium*; Peter Forbrig, Fabio Paternò, Annelise Mark Pejtersen; (Boston: Springer), pp. 33–46.

The main target of this paper is to draw attention to this problem by discussing recent research topics which address this problem using different approaches, and secondly, to point to problems which need to be investigated further. Hopefully, this will encourage more empirical studies and conceptualisations of the interaction among humans, their work and other variegated contexts and the technology used both within and across these contexts.

2. Background

HWID organises Annual Working Conferences with printed papers, discussions and varied forms of interactions and collaborations during two days of workshop activities. HWID'05, a Working Conference took place in Rome at INTERACT'05, the International Conference of Human Computer Interaction. The focus was on "Describing Users in Context". HWID'06 took place at the University of Maidera. The theme was "Synthesizing work analysis and design sketching", with a particular focus on how to read design sketches within different approaches. HWID'07 took place in Rio de Janeiro at INTERACT'07. The focus was on "Social, Organisational and Cultural aspects of Human-Work Interaction Design". The inspiration of this paper is based on fruitful dialogs during these HWID activities.

The subjects raised in this paper stem from the authors' analysis of papers written in conjunction with discussions that took place. These papers were written by researchers from around the world; the topics covered a variety of disciplines and theoretical approaches in human sciences: psychology, anthropology, sociology, information and media sciences; computer sciences and engineering.

Human-Work Interaction Design is a comprehensive approach in HCI, and in order to provide an easy understanding and to illustrate the coverage of this research topic, we developed the model in figure 1.

Figure 1 shows examples of the characteristics of humans and work domain contents and the interaction during their tasks and decision activities, individually or in collaboration. Analysis of users' work and life, as well as the design of computer-based information systems, has inspired the development of numerous theories, concepts, techniques and methods. Some have been widely adopted by practitioners; others are used mainly by researchers, and these are naturally part of Human-Work Interaction design research, and they will obviously influence the work and user analysis as well as the technology design. This explains the top box.

Environmental contexts, such as national, cultural, social and organizational factors, impact the way in which users interact with computers in their work and life to the same extent as the nature of the application domain, the tasks, and the users' skills and knowledge. The analysis and design of Human-Work Interaction will necessarily also include these contextual factors. As a consequence hereof, the bottom box of figure 1.

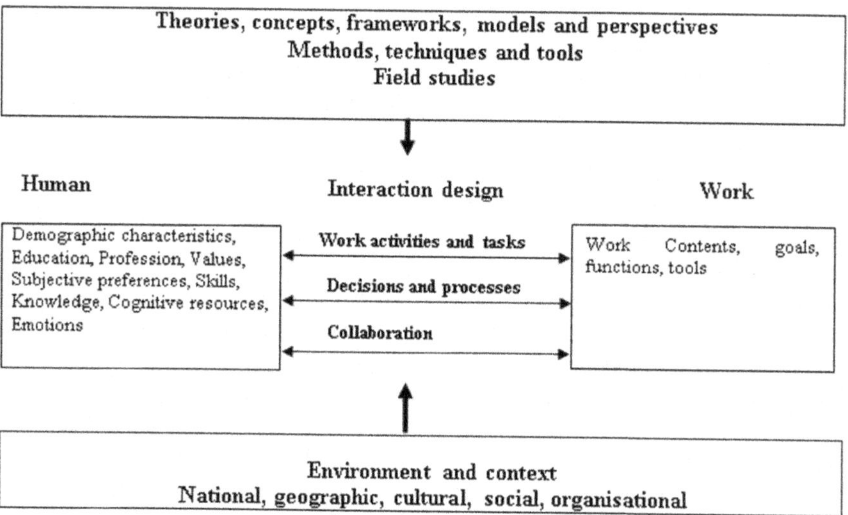

Figure 1 The Model of Human-Work Interaction Design

The analysis of the current HWID activities resulted in the identification of six main themes, which reflect those problems which the authors perceived to be of major concern in Human-Work Interaction Design:

Within Design processes

- Encouraging the dialogue between users and designers in the design process
- Bridging the HCI and Software Engineering gap by working with user requirements and collaboration in software development processes
- Supporting communication and design exploration through sketching

Within Work and User analysis

- Bridging the work analysis and interaction design gap through detailed case and field studies and action research projects empirical field studies.
- Rich contextual user descriptions, including methods to study unpredictable and opportunistic tasks
- Broadening the scope to Social, Organizational and Cultural aspects

Although this list presents these themes and problems as separated, they are of course intertwined and appear in different ways in many of the papers. Thus the following presentation of the papers within one of these headlines is of course exclusive for practical reasons, but as the reader will recognize, there are many overlapping themes and problems.

3. Dialogue in the design process, between users and designers

Design conceptualized as dialogue. Lopes 2006 provide a perspective on design as dialogue, consisting of a presentation of different definitions and different aspects of design, which could be argued as all being related to dialogue. Dialogue is considered in relation to objects, processes and disciplines of design. The author identifies some problems with the design-as-dialogue approach, mainly the complexity of the issue, and suggests a qualitative study that may help reveal ways to simplify and validate the approach.

Grounded theory to study users' responses. Nocera et al. 2005 suggest ways to support people's meetings and dialogues about their view of the world and their experience. They use grounded theory in the study of users' responses to an implementation of an ERP system in various countries; the authors investigate negotiation – as reconfiguration – between the roles of users and producers. The analysis shows very different attitudes toward the same systems when implemented in cultural diverse settings; it purports that making sense of the system in a particular work-context depends on cultural, organisational and individual preferences. These different attitudes and ways of use are particularly visible in breakdown situations; the authors argue for interaction between users and producers, and that producers should be able to observe and discuss users' breakdown situations, their frustration and workarounds.

Affinity Diagram for requirements elicitation. Bondarenko and Janssen 2005 use a different methodological approach. They use the Affinity Diagram method adapted from Hackos and Redish in the requirements elicitation process for the design of personal document management systems. Without losing the user's context and without requiring the reading of lengthy reports, this method helps structure large collections of mixed qualitative and quantitative data, and gives dynamic requirements (as opposed to static user profiles or task flows). However, the method as it is used per se results in abstraction of the requirements into a general level and hence results in difficulties in mapping the acquired results into system design.

Information acquisition using colleagues' verbal reports. Erlandsson and Jansson 2007. A new method for information acquisition called *collegial verbalisation* is explored using an empirical case study of vehicle operators being videotaped while driving a high-speed ferry, followed by some of their colleagues making verbal reports while watching this video data. These colleagues are very familiar with the driving task and the driver environment. The method is discussed in relation to the amount of information provided in general; the reliability of the data; and how it contributes to the detection of "buggy mental models" within the operators, and it is compared to more traditional forms of verbal reports. It is suggested that the method of collegial verbalisation may have combinatorial advantages that makes it more powerful as an analysis tool than the traditional forms of verbalisation, specifically if one wants to analyse work tasks that are dynamic and where the operators' behaviours are highly automated. However, more elaborate and systematic investigations must be conducted through experimental designs.

4 Bridging the HCI and Software Engineering gap

User interface model and requirement tool. España et al. 2006 look at the gap between HCI and Software Engineering (SE); while SE is supposed to be strong in specifying functional requirements, HCI is centred on defining user interaction at the appropriate level of abstraction. An abstract model of the user interface represented by a ConcurTaskTrees model is used to enrich the functional specification, and a new tool called RETO that aims at requirement engineering is presented. The adoption of such a framework is promising and future empirical studies will show if the model can be justified.

Activity Theory and software development process. Software development is intrinsically a collaborative activity. Based on an analysis of current literature and software, Lewandowski and Bourguin 2006 find that current Software Development Environments seldom provide true integrated collaboration between developers, rather they offer only sharing of material or communication support, and do not support the actual work process of software development. Further, the ability to tailor the development environment is an issue, as features for allowing external applications to be nested into the environment are lacking. Grounding their work on Activity Theory, the authors describe how the eclipse (open source software) has been extended to accommodate for some of these deficiencies; it will be interesting to follow these features being implemented in future development processes.

User interface patternsin specific contexts. Stanard and Wampler 2005 focus on richness multi-dimensionality of user descriptions, and discuss how design patterns until now have been close to traditional usability guidelines; thus, there is a need to make design patterns to better support interaction of specific contexts. User Interface (UI) patterns are presented as a way of defining, applying and evaluating the translation of cognitive and collaborative requirements into meaningful human computer interaction in the designed interface, and then through this provide input to the development process. The described case involves an airport control system, and the discussion of command and control systems. The patterns are useful not only for the provision of training and inspiration to solutions, but also for the reuse of patterns that have been quality assured in complex and risk environments, such as command and control systems. The authors argue for the need for hierarchies of patterns that are based on a specific application-domain or work-domain to enhance the work-performance.

Work style modelling In the same vein, Campos and Nunes 2005, 2006 combine Work Style modelling with Usage-Centered Design with the objective of designing and evaluating better design tools. They describe the richness in the human-work interaction by using a new method of work style modelling, which has been applied to the work-context of interaction designers (as well as to collaborative software design). The work style is described from a set of informally defined values, and the set of styles which has been shown apparent in the work-context, are then more formally depicted and evaluated using diagrams and metrics. By modelling users' work style, the focus is put on work transitions (from one style of work to the other) and the designed solution ability to support the current context and changes in these – within the same application. The authors raise the question of whether it is possible to use work style modelling in other fields to describe flows between contexts of use.

5 Sketching in Communication and design exploration

Collaborative design process. Craft and Cairns 2006 offer experiences with sketching in a design process for an information visualization tool. The objective of the system is to support communication between users with different backgrounds - between biologists and mathematicians. The authors present an in-depth analysis of the design process, showing that sketching as an integral part of a collaborative design process aids creativity, communication, and collaboration.

Representation of requirements based on Cognitive task analysis. Rozzi and Wong 2006 present a case study of how design sketching can be used as a technique for the representation of design requirements to help the creation of a common understanding between users, designers and software developers, during the development of a tool for supporting spatial-temporal reasoning in Air Traffic Control (ATC). The design process is based on a cognitive task analysis using the Critical Decision Method, relying on observation and video recordings as well as Contextual Inquiry interviews. The authors show how sketching was used to get insights into the design possibilities, but also find that spatial-temporal issues are difficult to illustrate with sketching techniques; thus, further work is needed.

Idea exploration and refinement of details. Orngreen 2006 reflects on what sketches are and on the use of design sketches when developing an e-learning platform for case-based learning. The author attempts to differentiate techniques that include sketches: rough hand drawn sketches - storyboards – prototypes, and how the emphasis changes from idea exploration to refinement of detail. The paper draws a distinction between a sketch as a design artefact that can stand alone and as part of a work process.

Reading design sketches using work analysis. Clemmensen 2006 investigates the role of design sketches in Interaction design and work analysis when designing a simple folder structure for e-learning software to be used for course administration at a higher education study programme. The author discusses how to conceptualize the process of reading design sketches using work analysis. The interface was evaluated using a think-aloud protocol, and was found to be less satisfactory than the earlier designs as it was 'long-winded'. This pointing to the need for future work on investigating the relation between the sketching techniques used and the design obtained in the development process.

Sketches to improve task performance. Although Pereira et al 2006 do not act in the space of IT, they adopt a human centred approach, illustrated with sketches, when looking to improve the performance of treadle pumps, to be used in developed countries. Similarly, Gaspar et al 2006, use annotations (words and sketches) to the photographs in their analysis and design studies when investigating ways to increase the amount of physical activity in the daily routine.

6 Bridging the gap between work analysis and interaction design

Cognitive work analysis and interface design. Upton and Doherty 2006 describe an approach to designing a visual application for a semiconductor manufacturing plant, which is seen as a complex, large-scale system requiring a structured design methodology. They present a design rationale supporting the explicit representation of hierarchies, the compatibility of views, and the use of contextual navigation. This design is derived from a cognitive work analysis, from which an Abstraction Decomposition Space (ADS) was made and the interface design was subsequently developed. The paper systematically describes the application of cognitive work analysis and the subsequent process of interface design, in an effort to bridge the design gap.

Future/vision seminars in action research. Based in user-centred and participatory design, Johansson and Sandblad 2006 investigate how a home care and help service organisation can be developed in order to be better prepared for future challenges. During their action research project, they used the future/vision seminar model, extended with assignments (such as: describe a day at work). The seminars resulted in the formulation of several scenarios, which again served as input to the design of a prototype.

Generic user interface for resource allocation. O'hargan and Guerlain 2006 provide a generic User Interface (UI) design for resource allocation problems. The UI is designed to support a person making resource allocation decisions (as opposed to purely automated decisions, often currently the case). They argue that their Resource Allocation Planning System (RAPS) can be adapted to several types of resource allocation domains. In future work it will be interesting to follow evaluations on whether or not it is capable of clearly supporting the work of people doing resource allocation.

Cognitive Task Analysis and Mapping analysis of team performance. Mapping analysis results into new designs in a multi-agent world. This is the focus of the proposal by McMorrow et al. 2005, who use cognitive task analysis to evaluate effective team performance in collaborative environments, such as air traffic management, in order to provide insights into how a technology becomes a 'team player'. A cognitive task analysis for effective team performance can help re-interpret the formal procedures often surrounding complex technological designs by negotiating among different perspectives and different meanings brought into the work environment.

Cognitive Work Analysis and train driver interfaces. Jansson, Olsson and Erlandsson 2007 conducted field studies on the improvement of existing train driver interfaces within the framework of cognitive work analysis (CWA) (Rasmussen, Pejtersen, Goodstein 1994 together with the method for collegial verbalisation which produces think-aloud protocols from video-recordings. The analyses show that the driver works in three rather separate time intervals: a long-range, a short-term and an immediate sense perspective. The driver switches between these while travelling between two stations. A prototype of a planning area of a driver interface was developed, making these switches and feed-forward planning possible. Early tests using the user centred design approach show that the planning area of the interface supports the feed-forward decision strategy. However, the driver group also made substantial changes in the design, indicating that UCSD is an efficient tool in order to capture user competencies, and to bridge the gap between analysis and design.

7. Rich contextual analysis of users

Multidimensional, multimedia portraits of users. Recognizing the need for a general format for user descriptions, Orngreen et al. 2005 present a theoretical focus on human beings as they are perceived by the designers of the technologies of the 21st century. They argue that today software developers use techniques and methods in software development that embed mono-cultural and mono-dimensional models in various contexts which in the future must be replaced by rich portraits of human beings. In continuation thereof, the same group of authors in Nielsen et al. 2006 argues that cultural embeddings are significant in relation to HCI because the cultural context is also embedded in the methodological framework, the techniques and the tools that we apply. The authors suggest a research program that aims at developing a theoretical framework supporting the creation of rich multimedia portraits of the human user of multimodal technologies Orngreen et al. 2005; the authors point to a theory of complementary positions that insists on solid accounts from all observer positions in relation to perspective, standpoint and focus Nielsen et al. 2006.

Activity theory, situated action and distributed cognition models. The need for different positions is also a theme in Kimani et al. 2005 who use activity theory, situated action and distributed cognition models to study the nature of tasks in real world, natural settings. Within the context of mobile computing, they focus on how supplementary tasks, such as interacting with the device, are performed while the user does another primary task. Unpredictable and opportunistic tasks can be studied with these beyond task-centric approaches in order to provide rich and complex descriptions of users in the mobile domain. Information Science is another domain, which requires discussion of current approaches to model and describe empirically the different kinds of contexts.

Information science Pejtersen et al. 2005 purport that we need not only an analysis of users' perceptual, cognitive, and social states, but also a deep understanding of how the users' contexts influence their interaction with artefacts such as a Digital Library. They propose that the problems raised within the information science field can provide a number of useful issues for discussion of the current approaches to describing users in context within the HCI field.

Critical Decision Method, Ethnography and Cognitive Work Analysis. Ham et al. 2005 present three case studies using three different methods, two for task-oriented design contexts (the Critical Decision Method and the Ethnography Method) and one for functional-oriented design contexts (the Cognitive Work Analysis Method, in particular the Abstraction Hierarchy). They argue that the critical decision method and the ethnography method provide useful and effective descriptions, enabling task-based design requirements in contexts of anticipated situations, while the abstraction hierarchy provides useful and effective descriptions in work domains of revolutionary designs for unanticipated situations. However, they miss an integrated method for obtaining information about user contexts, a method that is both task- and function-oriented.

The Activity Interview and Activity theory in HCI. Duignan, Noble and Biddle 2006 elaborate on their work on the activity interview based on cultural historical activity theory and in particular the activity checklist. The activity interview uses questions to get to an activity analysis as opposed to the abstract formulations of the activity checklist. The paper gives a thorough view of the activity theory relation to the HCI field and the activity list, and provides critical reflection of the list based on previous literature, as well as on personal experience. These discussions clearly bring forward issues for improvement at a very concrete level. In the future it will be interesting to follow the consequences that the activity interview has on design suggestions and how it can be seen in the resulting design. Further, it will be noteworthy to see whether the interview, as claimed, is appropriate for guiding the process of activity analysis, if performed by those who do not know activity theory or cultural historical activity theory.

8. Impact of social, organizational, cultural and historical factors

Avoiding cultural bias in usability tests. Clemmensen 2007 The CULTUSAB project is conducting an in-depth investigation of the key dimensions of culture that affect usability testing situations, including language, power distance, and cognitive style. All phases of the usability test are being evaluated for cultural impact, including planning, conducting, and reporting results. Special attention is being focused on subject-evaluator communication and cultural bias in the test design and structure of the user interface being tested. Experiments are being replicated in three countries: Denmark, India and China. The research will result in new testing methods and guidelines that increase the validity, by avoiding cultural bias, and allow for production of comparable results across countries.

Historical, national, and cultural factors in the work place. Rasmussen 2007 presents an empirical, qualitative study of Internet use in a National Film Archive in an Eastern European country. The purpose was to identify the use of and the attitude towards the Internet through field studies of individuals and organizations. The empirical study shows, that the staff at the archive only uses the Internet moderately in their work. It also shows that historical, national, and cultural factors can be used to explain the way people at work reacts to the new Internet technology. A cross-disciplinary study of the literature about Central and Eastern Europe made it possible to explain their behaviour and attitudes within a broader context.

A game based on cultural common sense. Anacleto Coutinho et al. 2007 argue that an effective educational process has to be instantiated in the local culture and that common sense knowledge represents culture. Common sense based games can be used to work on topics taught by teacher and can promote a meaningful learning, since the new knowledge (formal knowledge presented during classes) is related to pieces of knowledge already in the learners' cognitive structure (common sense knowledge). A common sense based game prototype to support the process of knowledge reinforcement of the content presented to students is presented. Teachers can set up a quiz game based on the Brazilian common sense knowledge. Preliminary analyses with users point out the potential for such approach.

9. What did we learn?

Obviously, a long list of specific and important problems can be derived from this research as described in each paper above. However, common issues are also addressed, which concern basic conditions of the HCI research.

While certain techniques and methods provide an integrated focus on analysis *and* design, most focus on either analysis or design. The strongest link between analysis and design is the general reliance on iteration as a way of developing products that fit the user needs and context, but within HWID other means and techniques have also been applied. Our papers and activities in the Working Group have operated on three levels:

- A field study level which involves an understanding of what actually goes on in a user environment.
- An applied level, which concentrates on methods and tools for analysis and design
- A theoretical level where academic disciplines have been selected to compensate for the shortcomings of single approaches when confronting the complexity of a design problem.

While experimental design of prototypes is a necessary component of the iterative process of work studies, design and evaluation, consistent conceptualisations between work analysis and application evaluation are needed to provide results that are valid beyond discrete experiments, and can be generalised to other application domains and contexts. In some papers the authors present a satisfactory result of the application of a specific approach to solve their defined problem, few are not successful, but the majority of papers present approaches to their problems which the authors find promising, although still problematic, or yet unresolved, because no evaluation has taken place, or because it is unknown whether the approach can be generalized beyond the application domain.

The diverse combination of the approaches have mostly been driven by a particular work domain context, which is why the concluding discussion of theoretical concepts and tools applied in empirical work and prototype designs often refer to further research for validation of these in other application domain.

It is obvious that further work needs to be done in evaluating the designs that have been made, not only as they work in everyday practice, but also in relating them back to the insights that were gained from the initial work analysis and interaction design phases; in this way it is possible to better inform the concepts, methods and techniques applied.

Figure 2 shows the human actors who interact with work domains during their collaborative tasks and decision activities. A variety of application domains are studied in HWID research papers and the humans who perform this work also spans many different characteristics. Within HWID many means and techniques have been applied to study particular design problems, in most papers not one, but several theories, concepts, techniques and methods from several scientific disciplines have been necessary.

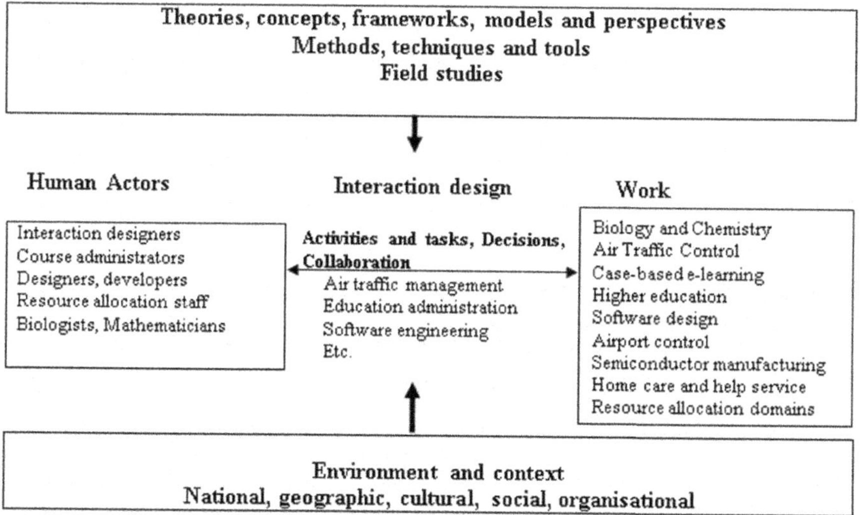

Figure 2 shows the application domains, the tasks and the users involved. The contextual factors in the buttom and the scientific approaches at the top.

There are domains where the work analysis shows that ICT are not the obvious solution by adopting a human centred approach, illustrated with sketches, when looking to improve the performance of treadle pumps, to be used in developed countries (Pereira 2006). Similarly, Gaspar et al 2006, use annotations (words and sketches) to the photographs in their analysis and design studies when investigating ways to increase the amount of physical activity in the daily routine.

Necessary in a global world, but still very emergent area in HCI with few research papers submitted, is to provide a better understanding of the complex interplay between individual, social, organizational, cultural, historical and national factors during the use of technology now and in the future.

10. Conclusion

Finally, we need to mention that although we have introduced many and most, not all, of the hot issues from our papers contributed by participants at the IFIP HWID Working Conferences, this paper's contribution is mostly to give an up to the minute account of research approaches within Human-Work Interaction Design. The informed reader will recognize that apart from the focus on work studies, many of the problems and approaches presented in this sketch are common for other HCI activities, although based on a relative small amount of papers, generalizations are not possible.

In spite of this limitation, it is our hope that the variety of challenges presented in this paper will inspire other researchers and readers to participate and contribute to a better understanding of the complexity involved. We hope this paper promotes the use of knowledge, concepts, methods and techniques that enables work and user studies and design experiments to procure a better apprehension of the complex interplay between individual, social, cultural and organisational contexts.

References

Anacleto Coutinho, Ferreira, Pereira (2007) "Promoting Culture Sensitive Education through a Common Sense Based Game" in Mark-Pejtersen, Clemmensen et al (Eds.) (2007): *Proceedings of the workshop: Social, Organisational and Cultural aspects of Human-Work Interaction Design, at the INTERACT 2007* conference, Rio, Brazil.

Bondarenko & Janssen (2005). "Affinity diagram method: Bringing users' context into the system design", Proceedings of the workshop: *Describing Users in Context – Perspectives on Human-Work Interaction*, the INTERACT 2005, Rome, Italy, p.34-37.

Campos & Nunes (2005). "A human-work interaction design approach by modeling the user's work styles", Proceedings of the workshop: *Describing Users in Context – Perspectives on Human-Work Interaction*, the INTERACT 2005, Rome, Italy, p.44-47.

Campos & Nunes (2006): "Principles and Practice of Work Style Modeling: Sketching Design Tools", Clemmensen, T., Campos, P., Ørngreen, R., Pejtersen, A. M., & Wong,

W. (Eds.). (2006). *Human work interaction design - designing for human work.:* Springer, New York. p. 203-220

Clemmensen (2006): "A simple design for a complex work domain", Clemmensen, T., Campos, P., Ørngreen, R., Pejtersen, A. M., & Wong, W. (Eds.). (2006). *Human work interaction design - designing for human work.*: Springer, New York. p. 221-240

Clemmensen (2007) "The Cultural Usability (CULTUSAB)" in Mark-Pejtersen, Clemmensen et al (Eds.) (2007): *Proceedings of the workshop: Social, Organisational and Cultural aspects of Human-Work Interaction Design, at the INTERACT 2007* conference, Rio, Brazil.

Craft & Cairns (2006): "Using Sketching to Aid the Collaborative Design of Information Visualisation Software", Clemmensen, T., Campos, P., Ørngreen, R., Pejtersen, A. M., & Wong, W. (Eds.). (2006). *Human work interaction design - designing for human work.*: Springer, New York. p. 103-122

Duignan, Noble & Biddle (2006): "Activity Theory for Design: From Checklist to Interview", Clemmensen, Campos, Ørngreen, Pejtersen, & Wong (Eds.). (2006). *Human work interaction design - designing for human work.*: Springer, N. Y.. p. 1-26

Erlandsson & Jansson (2007) "Collegial verbalisation – a case study on a new method on information acquisition" in Mark-Pejtersen, Clemmensen et al (Eds.) (2007): *Proceedings of the workshop: Social, Organisational and Cultural aspects of Human-Work Interaction Design, at the INTERACT 2007* conference, Rio, Brazil.

España, Pederiva, Ignacio Panach, Abrahão & Pastor (2006): "Linking requirements specification with interaction design and implementation", Clemmensen, T., Campos, P., Ørngreen, R., Pejtersen, A. M., & Wong, W. (Eds.). (2006). *Human work interaction design - designing for human work.:* Springer, New York. p. 123-134

Gaspar, Ventura, Pereira & Santos (2006): "Continuous fitness at home: Designing exercise equipment for the daily routine" Clemmensen, T., Campos, P., Ørngreen, R., Pejtersen, A. M., & Wong, W. (Eds.). (2006). *Human work interaction design - designing for human work.:* Springer, New York. p. 147-160

Ham, Wong & Amaldi (2005). "Comparison of three methods for analyzing human work in terms of design approaches", Proceedings of the workshop: *Describing Users in Context – Perspectives on Human-Work Interaction,* the INTERACT 2005, Rome, Italy, p7-11.

Jansson, Olsson & Erlandsson (2007): "Bridging the gap between analysis and design: Improving existing driver interfaces with tools from the framework of cognitive work analysis" in Mark-Pejtersen, Clemmensen et al (Eds.) (2007): *Proceedings of the workshop: Social, Organisational and Cultural aspects of Human-Work Interaction Design, at the INTERACT 2007* conference, Rio, Brazil.

Johansson & Sandblad (2006): "VIHO - Efficient IT Support in Home Care Services", Clemmensen, Campos, Ørngreen, Pejtersen, & Wong, (Eds.). (2006). *Human work interaction design - designing for human work.: Springer, New York.* p. 47-66

Kimani, Gabrielli & Catarci (2005)."Designing for primary tasks in mobile computing", Proceedings of the workshop: *Describing Users in Context – Perspectives on Human-Work Interaction,* at the INTERACT 2005, Rome, Italy, p.11-17.

Lewandowski & Bourguin (2006): "Improving collaboration support in software development activities" in Campos, Clemmensen, Orngreen, Wong, Mark-Pejtersen (2006) *the Pre-conference proceedings of HWID'06.* p. 33-45

Lopes (2006): "Design as Dialogue – a New Design Framework", Clemmensen, T., Campos, P., Ørngreen, R., Pejtersen, A. M., & Wong, W. (Eds.). (2006). *Human work interaction design - designing for human work.*: Springer, New York. p. 241-250

Mark-Pejtersen & Fidel (2005). "A Multi-Dimensional Approach to Describing Digital Library Users in Context", Proceedings of the workshop: *Describing Users in Context – Perspectives on Human-Work Interaction*, the INTERACT 2005, Rome, Italy, p.18-26.

Mark-Pejtersen, Clemmensen et al (2007): *Proceedings of the workshop: Social, Organisational and Cultural aspects of Human-Work Interaction Design, at the INTERACT 2007* conference, Rio, Brazil.

McMorrow, Amaldi & Boiardi (2005). "Approaches to designing for highly collaborative, distributed and safety-critical environments", Proceedings of the workshop: *Describing Users in Context – Perspectives on Human-Work Interaction*, the INTERACT 2005, Rome, Italy, p.38-43.

Nielsen, Yssing, Clemmensen, Orngreen, Nielsen, Levinsen (2006):" The human being in the 21st century– Design perspectives on the representation of users in IS" Clemmensen, Campos, Ørngreen, Pejtersen, A. M., & Wong, W. (Eds.). (2006). *Human work interaction design - designing for human work.*: Springer, New York. p. 93-102

Nocera, Dunckley & Hall (2005). "Reconfiguring producers and users through human-work interaction", Proceedings of the workshop: *Describing Users in Context – Perspectives on Human-Work Interaction*, at the INTERACT 2005, Rome, Italy, p.27-33.

O'Hargan & Guerlain (2006): Design of a Ressource Allocation Planning System" Clemmensen, Campos, Ørngreen, Pejtersen, & Wong, W. (Eds.). (2006). *Human work interaction design - designing for human work.: Springer, New York.* p. 67-92

Orngreen (2006): "The Design Sketching Process", Clemmensen, T., Campos, P., Ørngreen, R., Pejtersen, A. M., & Wong, W. (Eds.). (2006). *Human work interaction design - designing for human work.:* Springer, New York. p. 185-202

Orngreen, Clemmensen, Nielsen, Christiansen, Levinsen, Nielsen & Yssing (2005) "The Human Being in the 21st Century –Design perspectives on the representation of users in IS development, Proceedings of the workshop: *Describing Users in Contex – Perspectives on Human-Work Interaction*, at the INTERACT 2005, Rome, Italy, p.54-59.

Pereira, Malca, Gaspar & Ventura (2006): "Human Motion Analysis in Treadle Pump Devices", Clemmensen, Campos, Ørngreen., Pejtersen, & Wong (Eds.). (2006). *Human work interaction design - designing for human work.: Springer, New York.* p. 135-146

Rasmussen (2007), "Internet Use in Eastern Europe A Case Study" in Mark-Pejtersen, Clemmensen et al (Eds.) (2007): *Proceedings of the workshop: Social, Organisational and Cultural aspects of Human-Work Interaction Design, at the INTERACT 2007* conference, Rio, Brazil.

Rasmussen, Jens, Pejtersen,A.M. and Goodstein, L.P. (1994): Cognitive Systems Engineering. JohnWiley, London.

Rozzi & Wong (2006)" Design Sketching for Space and Time", Clemmensen, T., Campos, P., Ørngreen, R., Pejtersen, A. M., & Wong, W. (Eds.). (2006). *Human work interaction design - designing for human work.*: Springer, New York. p. 161-184

Stanard & Wampler (2005). "Work-centered user interface patterns" Proceedings of the workshop: *Describing Users in Context – Perspectives on Human-Work Interaction*, at the INTERACT 2005, Rome, Italy, p.48-53.

Upton & Doherty: "Visual Representation of Complex Information Structures in High Volume manufacturing", Clemmensen, T., Campos, P., Ørngreen, R., Pejtersen, A. M., & Wong, W. (Eds.). (2006). *Human work interaction design - designing for human work.*: Springer, New York. p. 27-46

Evaluating User Experience in Technology Pilots

Minna Isomursu

VTT, Finland, minna.isomursu@vtt.fi

Abstract: Computing devices and digital services have been moving rapidly from professional environments into the everyday life. This means that technology will influence the evolution of our everyday environment, including our physical surroundings, social encounters, and development of society. It becomes more and more important to evaluate the effects of technology in the realistic complex setting of everyday life in addition to controlled laboratory environments. In this paper, findings related to evaluating user experience in real-life trial conditions are summarized. The user experience evaluation methods are classified into four groups: (1) methods used before the pilot, (2) methods used during the pilot, (3) methods used immediately after the pilot and (4) follow-up studies. Each class bears their unique goals, possibilities and limitations for collecting user experience related data, and understanding it.

Keywords: user experience, piloting, trials

1. Introduction

This paper discusses the methods and problems related to collecting information about user experience evoked by ubiquitous applications and services in the context of technology pilots. Ubiquitous computing integrates technology with our everyday life and environment. The complexity and constraints of an everyday environment are difficult, if not impossible, to simulate in a laboratory environment. In situ evaluations are needed for evaluating new technological solution or technology based service in a setting with real users in real use environment (Consolvo, 2007). User experience evaluation can be done in this context for predicting or estimating what kind of user experience can be expected in real-world usage setting.

Capturing information about user experience is challenging, and real-life settings make it even more challenging as the everyday life context is complex and cannot be fully controlled. In an evaluation situation, challenges in capturing user experience occur on several levels and phases. Firstly, as human experience is

---

Please use the following format when citing this chapter:

Isomursu, M., 2008, in IFIP International Federation for Information Processing, Volume 272; *Human-Computer Interaction Symposium*; Peter Forbrig, Fabio Paternò, Annelise Mark Pejtersen; (Boston: Springer), pp. 47–52.

always subjective, the evaluation method should capture relevant parameters describing the user experience, which can then be recorded for analysis. Secondly, as user experience is dynamic (Forlizzi, 2000), it can change and evolve during the process of interaction. For example, at first the user can be happy and excited about the new product, but later become disappointed, sad or even angry if and when problems occur. Thus, user experience must be sampled several times during the use of the product, which often means organising long-term experiments. Thirdly, interpretation of captured data about user experience is difficult. For example, interpreting emotions from facial expressions captured on video has been an active and debated area of research for decades.

This paper discusses issues that arise when user experience is evaluated within the context of a technology pilot. Arranging a technology pilot requires that the technology under evaluation is mature enough that it can be used by real users in a realistic usage environment. However, the technology can be part of infrastructure or process that does not exists yet.

User experience is a subjective state. It does not have an objective reference, and therefore it cannot be objectively measured. An experience of one person cannot be experienced as such by another person. Therefore, systematic capture and analysis of user experience is very difficult. Furthermore, it is extremely difficult for humans to compare even their own experiences when they are separated by time. Human memory about experiences is utterly unreliable thus rendering our ability to recall past experiences so that we could compare them with other experiences, or describe them reliably after time has passed (Robinson, 2002). Also, our ability to predict our own experiences in a hypothetical or future setting is very limited (e.g. Gilbert, 1990). Therefore, the hypothesis of the research reported here is that the most reliable understanding of user experience can be achieved by: (a) evaluating user experience in a situation as close to actual realistic usage situation as possible to avoid the need for users to imagine or predict their experiences in a hypothetical situation, (b) collecting information and description of the experience at the time it happens to avoid the need to rely on the memories of the user in describing the experience, and (c) using the direct subjective information given by the person having the experience for defining and measuring the experience.

2. Research Setting

The work reported here has been done within the context of the SmartTouch (http://www.smarttouch.org) project, where ubiquitous mobile applications and services have been implemented and piloted within several application domains in field settings (e.g. Häikiö, 2007). The goal has been to evaluate new technological solutions with real users in real use environment. The hypothesis was that this can provide information about problems and issues that can be expected to occur in large-scale use in the real-world usage setting. Technology pilots aim at exposing the technology to real use under circumstances that can be observed and followed. In the research described here, user experience evaluation data collected through technology trials were used for evaluating the feasibility of technical construction, usability, value created for the end user, and the ethical issues related to adopting new technology.

3. Experiences From User Experience Evaluation In Technology Pilots

In this chapter, the user experience evaluation of technology pilot are divided into four phases: (1) before use evaluation, (2) during use evaluation, (3) after use evaluation, and (4) follow-up evaluation. Each evaluation phase has its specific goals, evaluation focus, and sets its own requirements for the evaluation methods. The user experience evaluation method should aim at not disturbing or changing the actual usage situation so that the actual user experience will change. However, the fact is that this is extremely difficult, as research has shown that mere measurement of a phenomenon has effects on the phenomenon itself (Morowitz, 1993 ; Heisenberg, 1927).

3.1 Before Use Evaluation

Before-use evaluations proved to be valuable especially for the following: (1) getting information about attitudes and expectations that are relevant for interpreting the results of the pilot, and (2) setting the baseline for the evaluation by describing and measuring the starting point so that improvements and changes introduced by the technology can be identified and measured.

Before pilot use it is the best time to evaluate the attitudes and expectations the users have towards issues that may be relevant for evaluating the results and impacts of the pilot. For example, in the elderly meal-ordering pilot, the attitudes towards the use of mobile phone proved to be a strong impacting factor towards the perceived usefulness of the service. Also, the pilot may result in attitude changes that can be identified only if the attitudes before and after the pilot can be measured and compared. In the SmartTouch project, information about attitudes and expectations were collected with contextual interviews when the sample size was small (under 20 users), and questionnaires when the experiment involved larger amounts of users. In some cases, questionnaires were printed on paper forms, and on others, they were implemented through a web-based survey tool. However, the experiences show that it is difficult to predict before the experiment what could be relevant parameters related to expectations and attitudes that would be needed for interpreting the results, as the values and attitudes of users often unfold only during the piloting. This could be solved by deeper user study concentrating on the values and attitudes of the users already before the pilot. Also, better models and methods for describing and modeling expectations and attitudes towards ubiquitous and persuasive technology would be needed.

The experiment often aims at improving or supporting the life of the pilot users in some way. For evaluating if improvement happens, the situation before the trial needs to be evaluated. For example, in the elderly meal-order case, one of the goals was to improve the satisfaction of the meal-ordering clients towards the meals offered. This could be done only by first evaluating the satisfaction towards meals before piloting, and then again after piloting. For evaluating improvement, it is crucial to identify right value creation parameters that are used for evaluation.

3.2 During Use

At the beginning of the piloting period, the pilot users are often introduced with the new technology under experimentation, and perhaps trained for using it. Observing the introduction and training situations allows a good opportunity for exploring the issues related to the adoption of the technology in question.

Collecting information about user experiences at the time they happen require in situ data collection methods (Consolvo, 2007) that can be applied during the use of technology. This means that the tools and methods used for collecting user experience data need to be integrated into the everyday practices of the pilot users, just as the technology under evaluation. Our experiences show that when the technology under evaluation is well integrated into the everyday practices of the user and therefore quite invisible, the user experience evaluation method may actually "steal the show" (Isomursu, 2007), if it is more visible and needs more attention and cognitive processing from the user, than the actual technology under evaluation. Humans are not very good at analyzing what actually caused an experience (Dutton, 1974), so it can be difficult for users to identify if the experience was caused by the technology under evaluation, or the user experience evaluation method (or any other event in the life of the pilot user).

During-use evaluation can focus not only on evaluating the user experience evoked by the technology under evaluation, but also how the technology affects the lives of its users. With ubiquitous technology, the technology is often invisible and therefore it is not designed to evoke experiences, but to help in daily tasks.

As the usage situations, including the physical and social environment, usage tasks, etc., may be very different between pilots, it can be necessary to integrate the experience collection method case by case into the pilot experiment. Also, automated compilation of activity logs make it possible to follow the actual usage patterns that have emerged during use.

3.3 *After Use*

At the end of the pilot use the users usually discontinue using the technology under piloting. This is a point where typically a feedback survey is performed. At this point, the users can report about their user experiences in the form of storytelling, and reflect on their experiences. However, as humans are naturally not very good in memorizing experiences, the limitations of after-use methods must be acknowledged.

After-use evaluation provides an opportunity to evaluate possible changes in attitudes of the users by comparing situations before and after use, and hearing the explanations of users for the possible attitude changes. Experiences indicate that the reply rates for questionnaires made after the pilot are higher than the questionnaires made before the pilot. One explanation might be that the pilot users feel they have more to contribute after the pilot as they are able to tell about their experiences by sharing stories. At this point of pilot experiments, users are familiar with the technology, its limitations and possibilities, and feel they better share the language and concepts used by technology developers and researchers. This can be exploited by combining after-use evaluation with brainstorming or other methods suitable for participatory design. Brainstorming sessions can be used for collecting user experience data, as improvement ideas and new scenario proposals often are loaded with user experience knowledge.

3.4 *Follow-up Study*

Follow-up studies are valuable in estimating the long-term effects of the experiment. The attitudes of trial users can fluctuate with time, and this does not necessarily end right after piloting. As the pilot use often provides the users with new possibilities to control their lives, depriving them from this feeling of control may have negative and even tragic effects (Schultz, 1978) that can be observed only after some time has elapsed after the experiment. The effects of the pilot experiment should be analyzed not only for collecting information about user experience, but also for evaluating the ethical issues related to the experiment. From the research point of view, it can be problematic to balance between avoiding negative impacts of the pilot and creating high-impact concepts. If the concept is found extremely valuable by the users and it is able to considerably contribute towards a higher quality of life for the pilot users, loosing the possibility to use technology after the pilot may have strong negative effect on the well-being of the pilot users. The negative psychological effect of loss can be much stronger than the positive effect achieved through pilot.

However, the goal of concept design is to create high-impact concepts that would be appreciated and valued by the users. Methods and examples for balancing between these two contradictory goals would be welcome.

4. Summary

This paper summarizes experiences about evaluating user experience with experimental pilots. The experiences reported in the paper have been collected in various experiments where the use of new technology has been evaluated in the everyday life of pilot users. In each experiment, a different set and combination of methods were used. For the purpose of this paper, the methods were classified into four classes depending on the point of time they have been used in the pilot process.

As it is very difficult, if not impossible, to fully understand and analyze the human experience, using several different methods in different phases of user experience evaluation can provide the designers and researchers with data that can be used to reveal details and characteristics of an experience from different viewpoints. Therefore, combining methods seems natural. Furthermore, as user experience is tightly context dependent, tailoring methods for each experimental context is probably also necessary. This means, that developing user experience capturing and evaluation methods can be as challenging as developing the technology under evaluation, as they both need to integrate and merge with the everyday lives of the pilot users.

References

Consolvo, S., Harrison, B., Smith, I., Chen, M., Everitt, K., Froehlich, J., Landay, J. Conducting In Situ Evaluations for and With Ubiquitous Computing Technologies. International Journal of Human-Computer Interaction. 22(1&2), 103-118 (2007)

Dutton, D., Aron A.: Some evidence for heightened sexual attraction under conditions of high anxiety. Journal of personality and social psychology, 30 (1974).

Forlizzi, J., Ford, S.: The building blocks of experience: an early framework for interaction designers. in Proceedings of the DIS 2000 Seminar, Communications of the ACM, New York, 419–423 (2000).

Heisenberg,W. Über den anschaulichen Inhalt der quantentheoretischen Kinematik und Mechanik, Zeitschrift für Physik, 43 (1927)

Häikiö, J., Isomursu, M., Matinmikko, T., Wallin, A., Ailisto, H., Huomo, T.: Touch-based user interface for elderly users. in Proceedings of MobileCHI, ACM Press (2007)

Isomursu, M., Tähti, M., Väinämö, S. and Kuutti, K.: Experimental Evaluation of Five Methods for Collecting Emotions in Field Settings with Mobile Applications. International Journal of Human Computer Studies. Elsevier. Volume 65 (Issue 4), 404 – 418 (2007)

Morwitz, V., Johnson, E., Schmittlein, D.: Does measuring intent change behavior. Journal of consumer research, 20, 1 453-469 (1993)

Robinson, M. Clore, G.: Belief and feeling: Evidence for an accessibility model of emotional self-report. Psychological bulletin, 128 (6), 934-960 (2002)

Schultz, R., Hanusa, B.: Long-term effects of control and predictability-enhancing interventions: Findings and ethical issues. Journal of personality and social psychology, 36 (1978)

Interface Model Elicitation from Textual Scenarios

Christophe Lemaigre, Josefina Guerrero García, and Jean Vanderdonckt

Belgian Laboratory of Computer-Human Interaction (BCHI)
Louvain School of Management (LSM), Université catholique de Louvain (UCL)
Place des Doyens, 1 – B-1348 Louvain-la-Neuve (Belgium)
E-mail: {christophe.lemaigre@, josefina.guerrero@student, jean.vanderdonckt@}uclouvain.be

Abstract: During the stage of system requirements gathering, model elicitation is aimed at identifying in textual scenarios model elements that are relevant for building a first version of models that will be further exploited in a model-driven engineering method. When multiple elements should be identified from multiple interrelated conceptual models, the complexity increases. Three method levels are successively examined to conduct model elicitation from textual scenarios for the purpose of conducting model-driven engineering of user interfaces: manual classification, dictionary-based classification, and nearly natural language understanding based on semantic tagging and chunk extraction. A model elicitation tool implementing these three levels is described and exemplified on a real-world case study for designing user interfaces to workflow information systems. The model elicitation process discussed in the case study involves several models: user, task, domain, organization, resources, and job.

Keywords: Model-driven engineering, requirements gathering, user interface development method, user task elicitation, workflow information systems.

1. Introduction

In recent years, there has been a lot of interest for scenario-based design (Rosson, 1997) and other forms of User-Centred Design (UCD) (Paterno, 1999) to initiate a development life cycle of User Interfaces (UI). Textual scenarios found in scenario-based design consist of informal but structured narrative descriptions of interaction sequences between the users and the interactive system, whether this system exists already or is simply envisioned. Scenarios have been proved (Rosson, 1997) to be a valuable tool to elicit, improve, and validate UI requirements.

Please use the following format when citing this chapter:

Lemaigre, C., García, J.G. and Vanderdonckt, J., 2008, in IFIP International Federation for Information Processing, Volume 272; *Human-Computer Interaction Symposium*; Peter Forbrig, Fabio Paternò, Annelise Mark Pejtersen; (Boston: Springer), pp. 53–66.

On the other hand, descriptions of the UI domain itself and the UI requirements are also expressed using conceptual models depicting either static (Tam, 1998) or dynamic (Fliedl, 2003) aspects of the interactive system. The models resulting from this process are supposed to raise the level of abstraction with respect to the implementation (Tam, 1998). The models are frequently expressed in a formal way so as to enable model reasoning. The process which ultimately leads to these descriptions, whether they are informal (such as scenarios) or (semi-)formal (such as models) is Requirement Engineering (RE) (Haumer, 1998).

Scenarios have the advantage to describe UI requirements from captured or imagined user interactions through concrete examples [8] of the user carrying out her task. This form is much more representative and evocative for an end user to validate UI requirements than models that are mainly used by software engineers. Models, e.g., domain models, user models, are expressed in a way that maximizes desirable properties such as completeness, consistency, and correctness (Vanderdonckt, 2005). But their expression is significantly less understandable for end users who are often in trouble of validating their UI requirements when they are confronted to models. Consequently, both types of descriptions, scenarios and models, are needed interchangeably in order to conduct a proper RE process that will effectively and efficiently feed the rest of the UI development life cycle. We introduce *model elicitation* as the general activity of transforming textual scenarios into models that are pertaining to the UI development.

The remainder of this paper is structured as follows: some related work is reported in Section 2. Three levels of model elicitation are defined in Section 3 and consistently described and discussed in the light of a model elicitation tool implementing these techniques. Section 4 will sum up the benefits and the shortcomings of the model elicitation techniques investigated so far and will present some future avenues for this work.

2. Related Work

Model elicitation consists of transforming scenarios into models so that they are usable in the rest of the development life cycle (Hemmecke, 2006), for instance by conducting a model-driven engineering method (Brasser, 2002; Vanderdonckt, 2005). *Model verbalization* (Jarrar, 2006) is the inverse process: it consists of transforming model elements into textual scenarios while preserving some quality properties (e.g., concision, consistency). Any model can be considered for this purpose: models found in HCI (e.g., task, user) or in RE (e.g., domain, organization). In (Bono, 1992), the system restates queries expressed on a domain model (here, an entity-relationship attribute model) into natural language expression.

As such, model elicitation is not new in Software Engineering (SE) (Fliedl, 2004, 2005b), but at least five significant works have been conducted in Human-Computer Interaction (HCI):

1. U-Tel (Tam, 1998) is a user-task elicitation software that enables designers to allocate elements of a textual scenarios into elements of three models: actions names (relevant to the task model), user classes (relevant to a user model), and objects names (relevant to a domain model). This allocation can be conducted manually or automatically.
2. ConcurTaskTrees Editor (Paterno, 1999) contains a module for task elicitation where designers copy task names found in a textual scenario and paste them in a graphical editor for representing a task model. Designers can then refine the task model, e.g., by specifying a task type, temporal relationships between tasks.
3. Similarly, T2T (Paris, 2002) is a tool for automatic acquisition of task elements (names and relationships) from textual documents such as manuals. Another version exists for the same purpose from a domain model (here, an object-oriented diagram) (Lu,1998) and for multiple heterogeneous sources (Lu, 2002).
4. Garland *et al.* (2001) present general software for gathering UI requirements from examples containing various elements that are relevant for different models, but models are not constructed per se.
5. Brasser & vander Linden (Brasser, 2002) developed a task elicitation system for the Isolde task modeling environment: based on a 25-state Augmented Transition Network (ATN) derived from written narratives, this system extracts two kinds of information: domain information (i.e., actors and objects) and procedural information (e.g., "when the user saves a file,…")

From these works, we observed the following shortcomings: some do not produce a genuine model at the end, for instance (Garland, 2001), some other produce model elements that are relevant to HCI, for instance (Lu, 1998; Paris, 2002), but only some model elements are derived (e.g., task names) or they mostly focus on task models whereas several models are typically found in HCI, not only the task model. When other models are considered, e.g., the user and the domain (Lu, 1998), only the names of the classes are captured. In this paper, we would like to capture all elements (i.e., classes, attributes, and relationships) of many interrelated models to inform the development. It is however fundamental that the task model is considered to initiate a full model-driven engineering life cycle (Clerckx, 2006; Paterno, 1999). Dynamo-AID (Clerckx, 2006) provides a distribution manager which distributes the sub-tasks of a task model to various computing platforms in the same physical environment, thus fostering a task-based approach for distributing UIs across locations of the physical environment. In the next section, an elicitation of UI model elements is provided according to three levels.

The three levels of model elicitation presented in this paper, i.e., manual classification, dictionary-base classification, and nearly-natural language classification, are presented in this order only for structuring purposes. This does not mean that the elicitation process should be conducted in that order. Indeed, one may desire eliciting model elements in a mostly automated way, then refine the classification manually. Or one may prefer first designating the most important model elements if they do not fit well from the identified ontology and then apply more automated techniques in order to propagate these manual classifications.

3. User Interface Model Elements Elicitation

In order to effectively support UI model elicitation, the model elements that are typically involved in the UI development life cycle should be considered. Figure 1 reproduces a simplified version of the ontology of these model elements that will be used throughout this paper: only classes and relationships are depicted here for concision, not their attributes and methods. The complete version of this ontology along with its definition and justification is detailed in (Guerrero, 2008).

We choose this ontology because it characterises the concepts used in the development life cycle of UIs for workflow systems, which are assumed to have the one of the largest coverage possible. Any other similar ontology could be used instead. In this ontology, tasks are organized into processes which are in turn ordered in a workflow. A job consists of a logical grouping of tasks, as we know them (Paterno, 1999). Jobs are usually assigned to organizational units (e.g., a department, a service) independently of the workers who are responsible to conduct these jobs. These workers are characterized thanks to the notion of user stereotype. But a same task could require other types of resources such as material resources (e.g., hardware, network) or immaterial resources (e.g., electricity, power). A task may manipulate objects that can invoke methods in order to ensure their role.

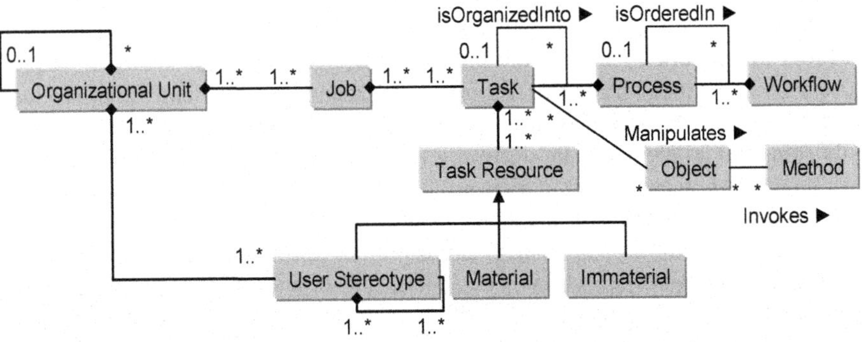

Figure 1. Simplified ontology of the model elements (Guerrero, 2008).

Figure 1 represents the conceptual coverage of model elements that will be subject to model elicitation techniques. This coverage is therefore larger than merely a task, an object, a user as observed today in the state of the art. In the next subsections, three progressively more sophisticated elicitation techniques based on this ontology will be described, motivated, and exemplified on a running textual scenario. This scenario explains the workflow for obtaining administrative documents in a town hall. The ordering of the three classification levels in the text is just a way to structure the article. Not an order the program user would comply in order to get a result.

3.1 Model Elicitation Level 1: Manual Classification

The UI designer is probably the most reliable person to identify in the textual scenario fragments that need to be elicited into model elements. Therefore, manual classification of model elements remains of high importance for flexibility, reliability, and speed. In a manual classification, any name that represents an instance of a model element belonging to the ontology can be manually selected, highlighted, and assigned to the corresponding concept, such as a task, a job, an organizational unit, etc. Consequently, all occurrences of this instance are automatically identified in the scenario and highlighted in the colour assigned to this concept. For instance, grey for an object, yellow for a user, red for an organizational unit, blue for a task. This colour coding scheme can be parameterized according to the designer's preferences.

Elicitation of a class. Any class belonging to the ontology can be manually classified according to the aforementioned technique. For example, "statement" is considered as an object in Figure 2 and is consequently assigned to the corresponding hierarchy in the related tab. Since a model element may appear in the scenario in multiple alternative forms (e.g., a plural form, a synonym), an alias mechanism enables designers to defines names that are considered equivalent to a previously defined one. For example, "statements" and "stated text" could be considered aliases of "statement".

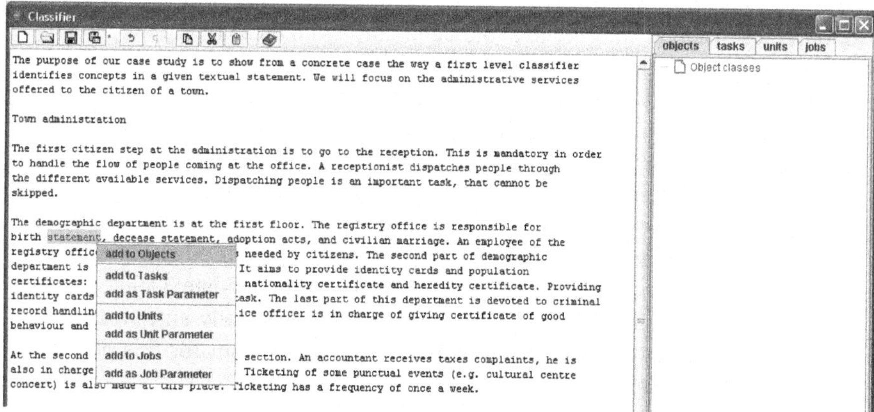

Figure 2. Elicitation of a class (here, an object) in manual classification.

In UCD, tasks, users, and objects are often considered as elements of primary interest. Therefore, it is likely that a designer will initiate the classification by identifying firstly tasks and related objects for instance. An object or a task could be of course elicited separately. In order to speed up this activity, the designer may directly associate a task to its related object when selected according to the same mechanism. All occurrences are highlighted similarly. Figure 3 illustrates this situation: a "birth statement" object is selected and a task "issuing" is attached to this object in order to create a complete task "issuing a birth statement". A special support exists for tasks: at any time, the designer may specify for a task a task type which belongs to one of the three following task types (Figure 4):

- A *predefined task type*: a taxonomy of task types (e.g., transmit, communicate, create, delete, duplicate) is made accessible for the designer to pick a name from, while a definition for each task type is displayed. This taxonomy consists of 15 basic task types that are decomposed into +/- 40 synonyms or sub-task types as used in the UsiXML User Interface Description Language (Vanderdonckt, 2005). This taxonomy has been established by relying on the Grounded Theory (Strauss, 1997), which means that it has been developed inductively from examining a corpus of data. In order to obtain this corpus, we have examined over time a series of interactive information systems and categorized the found task definitions into a corpus of task types that have been updated according to systematic deciphering scheme. Each predefined task type comes with a precise definition and scope, some synonyms, if any, and its decomposition into sub-tasks, if any. This taxonomy could be edited, e.g., by introducing some new task types, or by adding new synonyms to already existing task types.
- A *custom task name*: any non-predefined task name can be entered, such as "issuing" in Figure 3. In this way, any new task type that does not belong to the taxonomy may be introduced, such as task types for a particular domain of human activity. The custom task name is introduced mainly for specific case studies where one does not want necessarily to introduce a new task type in the taxonomy, for instance, in order to avoid deviations from these types.
- A *pattern of tasks*: a pattern of tasks is hereby referred to as any set of predefined task types and/or of custom task names. This should not be confused with a task pattern which is a pattern for task models. A pattern of tasks is aimed at gathering into one pattern a set of tasks that are typically, frequently carried out on an object. Instead of redefining every such task for an object, the pattern could be applied to the object, thus redefining the different tasks for this particular object. Such a set can be defined by the designer and reused at any time. For example, the pattern CRUD (acronym for Create, Read, Update, Delete), one of the most frequently applied patterns in SE, will automatically enter four predefined task types for a designated object and specialize them for this objects in order to avoid ambiguity.

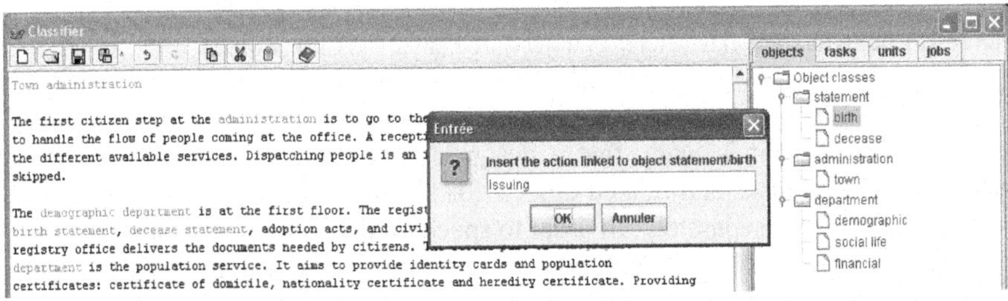

Figure 3. Assigning a task to a already defined object.

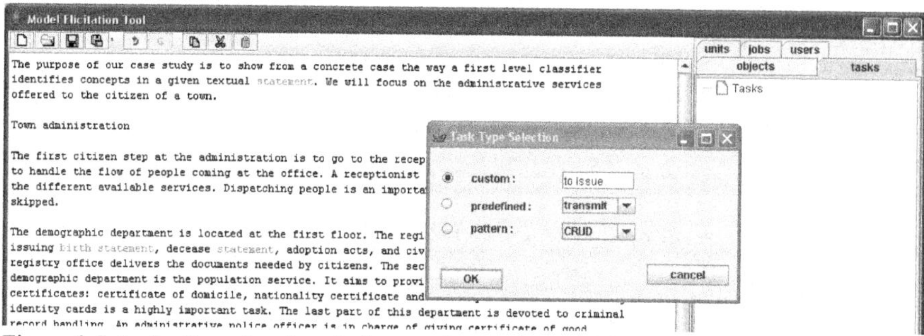

Figure 4. Introduction of various task types for a task model being elicited.

Elicitation of an attribute. The same technique is used in order to elicit an attribute of a class: either this attribute is predefined in the ontology (e.g., "frequency" to denote the frequency of a task) or a custom name can be manually entered. For example, in Figure 5, the designer has identified in the scenario the expression denoting the frequency of task and therefore elicits this attribute for the corresponding task (here, "ticketing"). The attribute is then represented as a facer of the corresponding task.

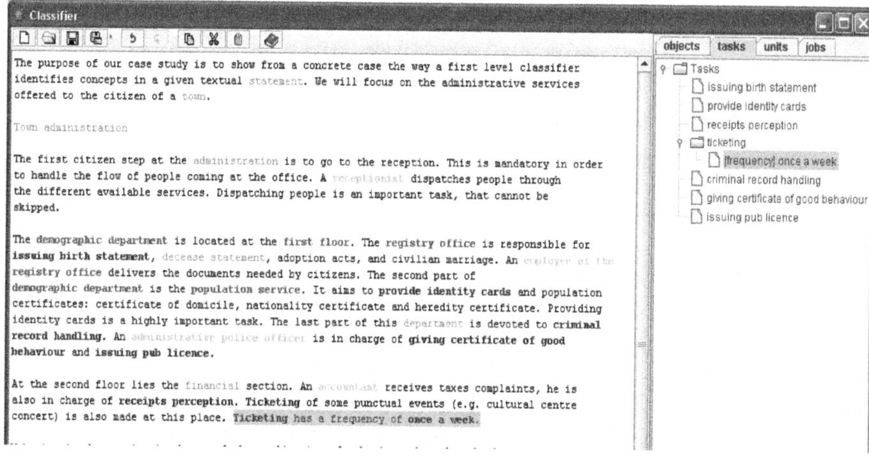

Figure 5. Elicitation of a predefined attribute for a task.

Figure 6 graphically depicts the three main steps for entering a custom name for an attribute, here an organizational unit. The procedure is similar for any other type of model attribute. The location of an organization unit is an attribute that does not belong to the ontology. Therefore, once such a parameter has been selected (Figure 6a), it is identified with a unique name (Figure 6b), and then included in the hierarchy (Figure 6c). Its value is then entered in the model as well. There is no underlying definition of data types supporting this action since it is considered rather informal at this stage of the development life cycle.

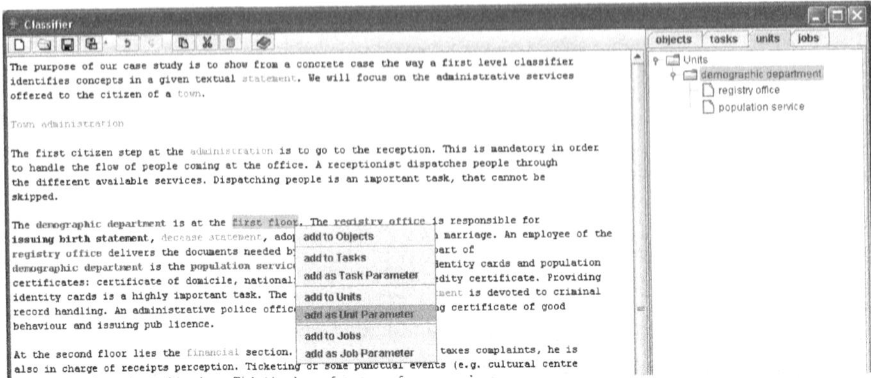

Figure 6a. Elicitation of a custom attribute for an organizational unit: selection.

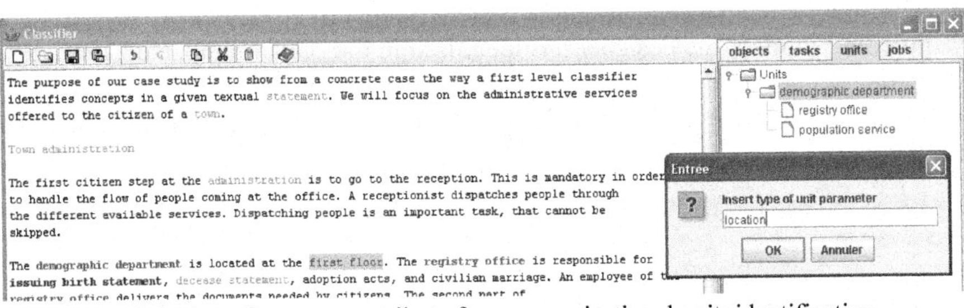

Figure 6b. Elicitation of a custom attribute for an organizational unit: identification.

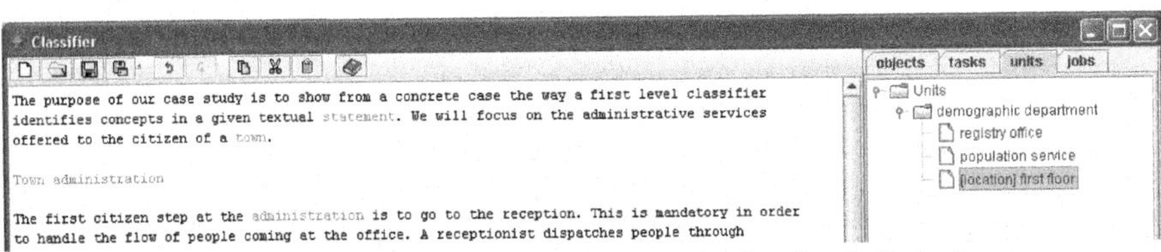

Figure 6c. Elicitation of a custom attribute for an organizational unit: inclusion.

Elicitation of a relationship. By using drag and drop, the designer can arrange model elements in their corresponding hierarchy in order to reflect the decomposition relationships of Figure 1. For example, a task is decomposed into sub-tasks, tasks are composed in a process, and processes are composed into a workflow. Or in the other way around, a workflow is decomposed into processes (e.g., business processes), which are in turn decomposed into tasks, to end up with sub-tasks. Apart from these decomposition relationships, only the "manipulates" relationship between a task and an object can be encoded in this level because it can be recorded thanks to the special support for tasks described above. For example in the right pane of Figure 3, the object "statement" is further refined into the two subclasses "birth statement" and "death statement", that automatically inherit from the attributes of the super-class.

3.2 Model Elicitation Level 2: Dictionary-based Classification

The model elicitation technique described in the previous sub-section, although flexible, reliable, and fast, represents a tedious task for the designer since it is likely to be repeated. Therefore, instead of manually designating in the textual scenario the fragments that are subject to model elicitation, these fragments could be automatically classified according to a dictionary of model terms. We distinguish two categories of dictionary:

1. *Generic dictionaries* contain fragments representing model elements that are supposed to be domain-independent (e.g., "a worker", "a manager", "a clerk" for a user model; "create", "read", "update", "delete" for a task model, etc.)
2. *Specific dictionaries* that contain fragments representing model elements that are domain-dependent (e.g., a "physician", "a pharmacist" in medicine for a user model; "administrate" for a task model, "physiology" for a domain model).

Each dictionary may contain predefined terms (like the task types) and aliases (e.g. plural, synonyms) in order to maximize the coverage of the automatic classification. In order to tailor this classification, two types of filters could be applied (Tam, 1998):

1. *Positive filters* force some model terms to be considered anyway, whatever the domain or the contexts of use are.
2. *Negative filters* prevent the automatic classification from classifying irrelevant terms, such as articles (e.g., "the", "a"), conjunctions (e.g., "with", "along"), etc.

The terms collected in such filters can be edited manually within any ASCII-compliant text editor. The advantage of this dictionary-based classification over the manual one is certainly its speed: in a very short amount of time, most terms belonging to the dictionaries, modulo their inclusion or rejection through the usage of filters, are classified. The most important drawback of this technique is that the identified terms are not necessarily located in the right place in their corresponding hierarchies. For example, a task hierarchy resulting from this process may consist of a one-level hierarchy of dozens of sub-tasks located in the same level without any relationships between them. In order to overcome this serious drawback a third level has been defined, which is the object of the next sub-section.

3.3 Model Elicitation Level 3: Towards Semantic Understanding

Different techniques exist that elicit model elements from textual scenarios, but so far they have never been applied in HCI to our knowledge: syntactic tagging (Fliedl, 2003), semantic tagging and chunk parsing (Fliedl, 2004). Genuine semantic understanding requires natural language understanding and processing, which is far beyond the scope of this work. What can be done however is to substitute a semantic understanding by a combination of syntactic and semantic tagging (Fliedl, 2004, 2005b) in order to recognize possible terms that express, depict, reveal model elements. For instance, a scenario sentence like "An accountant receives taxes complaints, but she is also in charge of receipts perception" should generate: a task "Receive taxes complaint", a task "charge of receipts perception", both being assigned to the user stereotype "Accountant", and a concurrency temporal operator between those two tasks because no specific term is included to designate how these tasks are actually carried out by an accountant. We may then assume the most general temporal operator, like a concurrency temporal operator. In order to reach this goal, this level attempts to identify possible terms in a syntactical structure (e.g., a set, a list, a sequence) that depicts a pattern for inferring for instance a task, another task with a temporal constraint, etc. For each model element, a table of possible terms involved in this pattern structure is maintained in accordance with the semantics defined in Figure 1. The parsing decides when to break any textual fragment (e.g., a sentence, a series of propositions that form a sentence) into separate model elements using both textual (e.g., periods, commas, colons, semi-colons) and lexical (e.g., "and", "or", "by", "to") cues.

Table 1. Possible terms incorporated in taggable expressions for a task model.

Concept	**Possible terms**
Frequency	Every day, daily, day by day, day after day, every Monday, every week, weekly, monthly, each month, each year, yearly, two (#) times a day, each hour, two (#) times per month (day, year, hour), two (#) days (weeks, months) a week (month, year), occasionally, from time to time, every other day, on alternate days, each two (#) days (weeks, months, years, hours)
Importance	Very important, low, high, regular
Structuration level	Low, high, regular
Complexity level	Low, high, regular, trivial, very complex, simple
Criticity	Low, regular, high, very critic
Centrality	Low, high, regular, very central, minor, peripheral
Termination value	End, finish, last, final, finally, lastly, endings
Task type	Communicate, convey, transmit, call, acknowledge, respond/answer, suggest, direct, instruct, request, create, input/encode/enter associate, name, group, introduce, insert, (new), assemble, aggregate, overlay (cover), add, delete, eliminate, remove/cut, ungroup, disassociate, duplicate, clone, twin, reproduce, copy, filter, segregate, set aside, mediate, analyze, synthesize, compare, evaluate, decide, modify, change alter, transform, turning, rename, segregate, resize, move, relocate, navigation, Go/To, perceive, acquire/detect/search for/scan/extract, identify / discriminate / recognize, locate, examine, monitor, scan, detect, Reinitialize, wipe out, clear, erase, select/choose, pick, start, initiate/trigger, play, search, active, execute, function, record, purchase, stop, end, finish, exit, suspend, complete, terminate, cancel, toggle, activate, deactivate, switch
Task item	Collection, container, element, operation

Table 2. Possible terms incorporated in taggable expressions for a task relationship.

Relationship	Possible terms
Sequence	Then…after; following; first, second…
Parallel	And; at same time; in any order; in parallel; jointly; concurrently
Conditional	If… then … else; either… or; in case of … otherwise
Iterative	Each time
Suspend / resume	Stop; suspend; discontinue; cease

On the one hand, this pattern matching scheme is syntactical because it is only based on detecting a particular combination of terms. On the other hand, those terms are assumed to reflect the precise semantics defined in the ontology. But we cannot say that this is a true semantic understanding anyway. Table 1 shows some excerpts of possible terms related to the concept of task, along with its attributes, while Table 2 shows some possible terms for detecting possible temporal relationships between tasks; these values are the result of the exploration of existing literature. This pattern matching can be executed automatically or under the control of the designer who validates each matching one by one.

The reserved names for model elements (e.g., task, the task attributes, and the temporal operators between the tasks) are read from the XML schema definition of the underlying User Interface Description Language (UIDL), which is UsiXML (Vanderdonckt, 2005) in this case. This XSD file can be downloaded from http://www.usixml.org/ index.php?mod=pages&id=5.

3.4 After Model Elicitation

The main goal of model elicitation is then to handle the textual statement from the beginning to the end and to ensure that all textual fragments that should be transformed into model elements are indeed elicited. In particular, the graphical highlighting in colours allows designers to quickly identify to which model type the element is relevant and to check in the end that the complete scenario has been exhausted, that no term remains unconsidered. In this way, they can check whether main model properties are addressed in an informal way, such as, but not limited to those model properties that are summarized in Table 3.

It seems of course impossible to automatically check these model properties at this level since only textual fragments are considered, even if they are linked with the ontology. However, this table may serve as a check list to ensure that the resulting models are the least incomplete, inconsistent, incorrect, etc. as possible.

After performing the elicitation of model elements according to any of the three aforementioned techniques, the model elicitation tool can export at any time the results in UsiXML files for the whole set of models or for any particular combination (e.g., only the tasks with the users or only the tasks with their related objects).

This file can then be imported in any other UsiXML-compliant editor in order to proceed with the remainder of the development life cycle. Several tools are candidates for this purpose (Vanderdonckt, 2005):

- IdealXML enables designers to graphically edit respectively the task and the domain models, in particular to automatically generate an Abstract UI.
- FlowiXML enables designers to edit the task, job, and organizational unit models in order to proceed with user interfaces for workflow information systems. Per se, it does not edit the domain model however. It is mainly targeted towards editing models that are involved in workflow information systems.
- Any general-purpose tool for applying model-to-model or model-to-code transformations, in particular any software that supports solving the mapping problem [20] between various models.

Table 3. Desirable quality properties of a model.

Property	Definition
Completeness	Ability of the model to abstract *all* real world aspects of interest via appropriate concepts and relations
Graphical completeness	Ability of the model to represent *all* real world aspects of interest via appropriate graphical representation of the concepts and relations
Consistency	Ability of the model to produce an abstraction in a way that reproduces the behaviour of the real world aspect of interest in the same way throughout the model and that preserves this behaviour throughout any manipulation of the model.
Correction	Ability of the model to produce an abstraction in a way that correctly reproduces the behaviour of the real world aspect of interest
Expressiveness	Ability of the model to express via an abstraction any real world aspect of interest
Concision	Ability of the model to produce concise, compact abstractions to abstract real world aspects of interest
Separability	Ability of models to univocally classify any abstraction of a real world aspect of interest into one single model (based on the principle of *Separation of Concerns*)
Correlability	Ability of models to univocally and unambiguously establish relations between models to represent a real world aspect of interest
Integrability	Ability of models to concentrate and integrate abstractions of real world aspects of interest into a single model or a small list of them.

4. Conclusion

In this paper, we have investigated three different techniques for eliciting model elements from fragments found in a textual scenario in order to support activities of scenario-based design. These three techniques are progressively more advanced in terms of consideration of the possible terms found in the scenario: from purely manual syntactical classification until ontology-based pseudo-semantic understanding. These three levels can be used in combination. Beyond the automated classification of terms into the respective models that are compatible with the ontology, the model elicitation tool provides editing facilities within a same model and across models of this ontology. In order to support other models or other variations of the same model (e.g., a different task model or more attributes for the same task model), one may need to incorporate these definitions in the ontology. As empirical validation is an important component in understanding the capacity and limitations of the model elicitation tool, a series of case studies has been developed.

Its main advantage relies in its capability of supporting designers in identifying text fragments that should be considered for model elicitation and in helping them to informally check some desirable model properties.

Its main drawback today is the lack of graphical visualisation of inter-model relationships or intra-model relationships others than merely decomposition relationships (represented implicitly in the respective hierarchies). Advanced visualisation techniques, such as carrousel visualisation, may be considered. For the moment, these relationships are only collected as an entry in a table that can be further edited. In the near future, we would like to refine the level 3-technique in terms of possible combinations of terms in an expression to be subject to semantic pattern matching.

Acknowledgments. We gratefully acknowledge the support of the SIMILAR network of excellence (http://www.similar.cc), the European research task force creating human-machine interfaces similar to human-human communication of the European Sixth Framework Programme (FP6-2002-IST1-507609) and the CONACYT program (www.conacyt.mx) supported by the Mexican government. We also thank the anonymous reviewers for their constructive comments.

References

Bono, G., and Ficorelli, P.: Natural Language Restatement of Queries Expressed in a Graphical Language. In: Proc. of the 11th Int. Conf. on the Entity-Relationship Approach ERA'92 (Karlsruhe, October 7-9, 1992), Lecture Notes in Computer Science, Vol. 645, pp. 357-373, Springer, Heidelberg (1992).

Brasser, M., and Vander Linden, K.: Automatically Elicitating Task Models from Written Task Narratives. In: Proc. of the 4th Int. Conf. on Computer-Aided Design of User Interfaces CADUI'2002 (Valenciennes, 15-17 May 2002), pp. 83-90, Kluwer Academic Publishers, Dordrecht (2002).

Clerckx, T., Vandervelpen, Ch., Luyten, K., and Coninx, K.: A Task Driven User Interface Architecture for Ambient Intelligent Environments. In: Proc. of 10th ACM Int. Conf. on Intelligent User Interfaces IUI'2006 (Sydney, January 29-February 1, 2006), pp. 309-311, ACM Press, New York (2006).

Fliedl, G., Kop, Ch., Mayr, H., Hölbling, M., Horn, Th., Weber, G., and Winkler, Ch.: Extended Tagging and Interpretation Tools for Mapping Requirements Texts to Conceptual (Predesign) Models. In: Proc. of 10th Int. Conf. on Applications of Natural Language to Information Systems NLDB'2005 (Alicante, June 15-17, 2005), Lecture Notes in Computer Science, Vol. 3513, pp. 173-180, Springer, Heidelberg (2005).

Fliedl, G., Kop, Ch., and Mayr, H.: From textual scenarios to a conceptual schema. Data Knowledge Engineering, 55(1), 20-37 (2005).

Fliedl, G., Kop, Ch., Mayr, H., Winkler, Ch., Weber, G., and Salbrechter, A.: Semantic Tagging and Chunk-Parsing in Dynamic Modeling. In: Proc. of 9th Int. Conf. on Applications of Natural Languages to Information Systems NLDB'2004 (Salford, June 23-25, 2004), Lecture Notes in Computer Science, Vol. 3136, pp. 421-426, Springer, Heidelberg (2004).

Fliedl, G., Kop, C., and Mayr, H.: From Scenarios to KCPM Dynamic Schemas: Aspects of Automatic Mapping. In: Proc. of 8th Int. Conf. on Applications of Natural Language to Information Systems NLDB'2003 (Burg, June 2003), Lecture Notes in Informatics, Vol. 29, pp. 91-105, Gesellschaft für Informatik, Bonn (2003).

Garland, A., Ryall, K., and Rich, Ch.: Learning hierarchical task models by defining and refining examples. In: Proc. of the 1st Int. Conf. on Knowledge Capture K-CAP'2001 (Victoria, October 21-23, 2001), pp. 44-51, ACM Press, New York (2001).

Guerrero, J., and Vanderdonckt, J.: FlowiXML: a Step towards Designing Workflow Management Systems, Journal of Web Engineering, 4(2), 163-182 (2008).

Haumer, P., Pohl, K., and Weidenhaupt, K.: Requirements Elicitation and Validation with Real World Scenes, IEEE Transactions on Software Engineering, 24(12), 1036-1054 (1998).

Hemmecke, J., and Stary, Ch.: The Tacit Dimension of User Tasks: Elicitation and Contextual Representation. In: Proc. of 5th Int. Workshop on Task Models and Diagrams for User Interface Design TAMODIA'2006 (Hasselt, October 23-24, 2006), Lecture Notes in Comp. Science, Vol. 4385, pp. 308-323, Springer, Heidelberg (2006).

Jarrar, M., Keet, M., and Dongilli, P.: Multilingual verbalization of ORM conceptual models and axiomatized ontologies. Technical report. STARLab. (Available via Vrije Universiteit, 2006). http://www.starlab.vub.ac.be/staff/mustafa/publications/[JKD06a].pdf. Accessed 14 April 2008.

Lu, S., Paris, C., and Vander Linden, K.: Computer Aided Task Model Acquisition From Heterogeneous Sources. In: D. Guozhong (Ed.), Proc. of 5th Asia Pacific Conference on Computer Human Interaction APCHI'2002 (Beijing, November 1-4, 2002), pp. 878-886, Science Press, Beijing (2002).

Lu, S., Paris, C., and Vander Linden, K.: Toward the Automatic Construction of Task Models from Object-Oriented Diagrams. In: Proc. of the IFIP TC2/TC13 WG2.7/ WG13.4 7th Working Conf. on Engineering for Human-Computer Interaction EHCI'98 (Heraklion, September 14-18, 1998), pp. 169-189, IFIP Conference Proceedings, Kluwer (1999).

Paris, C., and Vander Linden, K., Lu, S.: Automated knowledge acquisition for instructional text generation. In: Proc. of the 20th Annual Int. Conf. on Computer documentation SIGDOC'2002 (Toronto, October 20-23, 2002), pp. 142-151, ACM Press, New York (2002).

Paterno, F., and Mancini, C.: Developing task models from informal scenarios. In: Proc. of ACM Conf. on Human Aspects in Computing Systems CHI'99 (Pittsburgh, May 15-20, 1999), ACM Press, New York (1999).

Rosson, M.B., Carroll, J. M.: Scenario-based Design. In: A. Sears, J.A. Jacko (Eds.), The human-computer interaction handbook: fundamentals, evolving technologies, and emerging applications, CRC Press (2007).

Strauss A.L., Corbin, J.: Grounded Theory in Practice, Sage, London (1997).

Tam, R., Maulsby, D., Puerta, A.: U-TEL: A Tool for Eliciting User Task Models from Domain Experts. In: Proc. of ACM Int. Conf. on Intelligent User Interfaces IUI'1998 (San Francisco, January 6-9, 1998), pp. 77-80, ACM Press, New York (1998).

Vanderdonckt, J.: A MDA-Compliant Environment for Developing User Interfaces of Information Systems. In: Proc. of 17th Conf. on Advanced Information Systems Engineering CAiSE'05 (Porto, June 13-17, 2005), Lecture Notes in Computer Science, Vol. 3520, pp. 16-31, Springer, Heidelberg (2005).

Virtual Fixtures for Secondary Tasks

G. Lefemine[1], G. Pedrini[2], C. Secchi[3], F. Tesauri[4], and S. Marzani[5]

[1] DISMI, University of Modena and Reggio Emilia, via Amendola 2, Morselli Building, 42100 Reggio Emilia, Italy, lefemine.gianvito.34608@unimore.it
[2] DISMI, University of Modena and Reggio Emilia, via Amendola 2, Morselli Building, 42100 Reggio Emilia, Italy, pedrini.guido.39843
[3] DISMI, University of Modena and Reggio Emilia, via Amendola 2, Morselli Building, 42100 Reggio Emilia, Italy, cristian.secchi@unimore.it
[4] DISMI, University of Modena and Reggio Emilia, via Amendola 2, Morselli Building, 42100 Reggio Emilia, Italy, francesco.tesauri@unimore.it
[5] DISMI, University of Modena and Reggio Emilia, via Amendola 2, Morselli Building, 42100 Reggio Emilia, Italy, stefano.marzani@unimore.it

Abstract: The insertion of data in personal devices (e.g. mobile phones, GPS devices) tends to distract us from the primary task (e.g driving) that we are executing because of the necessity of deviating our visual attention to a secondary task. In this work we have tested the benefits introduced by the haptic feedback as a facility for a very common secondary tasks, namely the insertion of strings in an input device. Experiments demonstrate that the presence of virtual fixtures improves performances during input tasks and decreases the distraction of the user from the primary task.

Keywords: haptic, virtual fixtures, affordances, secondary tasks, input task

1. Introduction

In our daily life, we are surrounded by personal devices (e.g. mobile phones, pocket PCs, GPS) that require our attention both for delivering their outputs and for giving them some inputs (e.g. writing an SMS, inserting a destination in a GPS device). Usually, what we are doing with these personal devices is not our main activity and, therefore, these requests of attention increase the amount of distraction from our primary activity. For example, distraction is a relevant concern about in-vehicle information systems (IVIS): drivers must divert part of their attentive resources from the driving task (*primary task*), in order to perform input actions and to receive and understand the system output (*secondary task*), see (Young, 2003). So far, several researches have investigated driver's distraction in order to isolate factors affecting driving performance and to develop

Please use the following format when citing this chapter:

Lefemine, G., et al., 2008, in IFIP International Federation for Information Processing, Volume 272; *Human-Computer Interaction Symposium*; Peter Forbrig, Fabio Paternò, Annelise Mark Pejtersen; (Boston: Springer), pp. 67–81.

distraction-mitigating IVIS, see for example (Donmez, 2004). Still, most of these studies mainly addressed to the second half of the problem, namely how the system informative output should be delivered to the driver in order to minimize the distraction impact. Only few studies (e.g. Nowakowski, 2000) have addressed the problem of defining which input strategies for the secondary task could fit the driving context at best, that is, which kind of device could allow users to safely perform input tasks while driving. At present haptic technologies seem to be the most promising way to achieve the result of minimizing distraction on a secondary input task in a driving context. Haptic feedback can be exploited to give to the input device a higher affordance (Gibson, 1977) and, consequently, to make its use easier on behalf of the driver. A first attempt in this direction has been done by BMW, as reported in (Bengler, 2002). The increase of distraction from the primary task is due to the fact that the same sensorial channel (e.g. vision in the driving context) is significantly required for the completion both of the primary (e.g. driving) and of the secondary task (e.g. inserting a destination on a GPS device). In case two different sensorial channels are involved for the completion of the primary and of the secondary tasks, the distraction from the primary task and the completion time of the secondary task should decrease. Very often, the sensorial channel requested for completing a primary tasks is vision. The goal of this work is to evaluate the benefits of the use of the haptic feedback as the main sensorial channel involved in the insertion of data in input devices. We have developed a prototype of a virtual keyboard input system over which two input strategies have been tested and compared. The first strategy basically consists of a virtual keyboard over which the user moves a pointer; each letter is selected by taking the pointer over it and clicking; the visual attention of the user is required for the insertion process. The second strategy endows the keyboard with a set of virtual fixtures (see, for example, Rosenberg, 1993; Bettini, 2004; Payandeh, 2002; Nolin, 2003) which are activated following a search algorithm called SAPETS (Search Algorithm for Possible Endings of Typed Symbols). The goal of SAPETS is to activate, depending on the letters already selected and on a set of words contained in a database, a set of fixtures that suggest to the user the possible completions of the word he/she is introducing. The logic behind this strategy is basically that of reducing the load on the users' visual attention. This mainly happens in two ways: on the one hand, visual scanning among keys is reduced by presenting visual cues; on the other hand, movements to be performed are haptically guided, thus minimizing the need for fine adjustment. Several experiments have been conducted in order to assess whether the presence of virtual fixtures provides a significant benefit for the user both in case the input task is the only one to perform and in case the user has to draw attention to a primary task, being the input task a secondary task. The paper is organized as follows: in Sec. 2 we provide a description of the experimental setup and of the SAPETS search algorithm used for activating the fixtures over the virtual keyboard. In Sec. 3 and Sec. 4 we provide the results of two sets of experiments conducted for evaluating the benefits introduced by the virtual fixtures. In the first case, the only task that has to be carried on by the user is the introduction of words through the input device. In the second case the user has to pay attention to a primary task and to insert words through the virtual keyboard. Finally, in Sec. 5, some concluding remarks are reported and some future work is addressed.

2. The Experimental Setup

The aim of this section is to describe the prototype of the input interface that has been developed and the algorithm that is used for enabling the virtual fixtures over the keyboard. The prototype is made up of an haptic device through which the user can interact with a virtual environment representing the input device. We have used a Phantom Omni haptic device by Sensable Technologies and we have implemented the virtual input device in a Matlab/Simulink environment using the Virtual Reality Toolbox. The interconnection between the Phantom Omni and the virtual environment has been implemented by using the Handshake Prosense Virtual Touch toolbox. In this way, the overall application can be developed using Simulink. In fact it is possible to interact both with the virtual environment and with the haptic device by means of Simulink blocks and the development and the maintenance of the overall interface is very quick and intuitive. The virtual environment that has been designed represents a QWERTY keyboard, see Figure 1.

The virtual keyboard enables typing of characters strings and has several typing functions and options (e.g. confirm a string, cancel a character). Each key is represented by the correspondent letter and it is enclosed neither in a visual nor in a physical container (unlike the physical keys of the normal keyboards). The user moves, through the Phantom, a pointer that is used for interacting with the keyboard. The selection of a key is made by moving the pointer on the desired character and by pressing the white button that is placed on the body of the stylus. Since the virtual keyboard doesn't have shaped keys, a haptic elastic potential well has been implemented to make keys selection easier: when the cursor is close to the letter (within a circle with radius of 10 mm and centered in the center of the rectangle in which the letter can be contained), it is elastically attracted to the center.

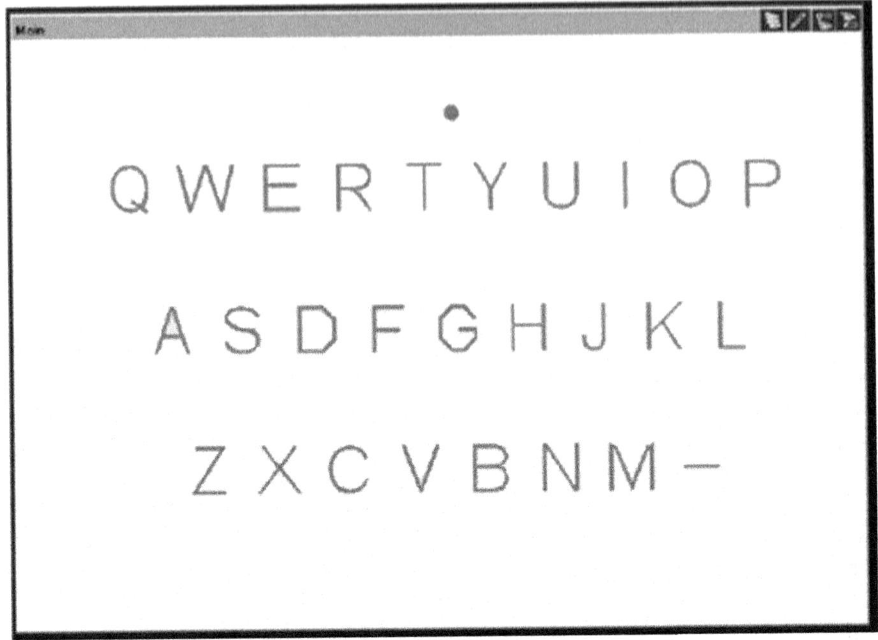

Figure 1 The virtual keyboard layout

In this way, the user can just roughly approach the pointer to the letter and, then, the potential well will take care of bringing the pointer exactly in correspondence of the letter. The user can select the letter over which the cursor is placed by pressing the white button on the stylus. In order to change letter, it is necessary to apply a little force for defeating the elastic force imposed by the potential well around the selected letter and to move the pointer to the next desired point. In this version of the keyboard no haptic facility for inserting words has been introduced. A second virtual keyboard, obtained by endowing the one just described with virtual fixtures for helping the user during the word insertion process, has been developed. For this application, a virtual fixture is a track that joins a pair of letters of the virtual keyboard. When the user moves the pointer along the track, a virtual force constrains the user to keep on moving the pointer along the track. The virtual force is local and, therefore, the user can take the pointer in an off-track position by applying a force that is sufficiently high. The virtual fixtures activation is governed by the SAPETS (Search Algorithm for Possible Endings of Typed Symbols) algorithm. Loosely speaking, the main idea behind this algorithm is to activate the virtual fixtures, namely some preferred directions the user should drive the pointer along, similarly to how the T9 software, developed by Tegic Communications Inc. and present on the most part of mobile phones currently on the market, suggests possible completions of the words while typing SMS. A set of words is initially stored in a database. Each time that the user inserts a letter, the SAPETS algorithm is activated. It consists of two stages: the first one searches for all the words that may possibly complete the typing, while the second one looks for all the possible letters that may follow the typed characters. As a result of the SAPETS algorithm, in the second version of the keyboard, after the insertion of each character, visual tracks appears, joining the character to those which could possibly follow it, according to the vocabulary stored in the database; around such tracks is implemented an elastic potential well whose role is to keep the pointer on the track (as proposed in Nolin, 2003; Secchi, 2006). The user can force the pointer to go off the track by applying the amount of force necessary for escaping from the potential well. The virtual fixtures have been placed on a geometric plane which is layered some millimeters over the keyboard plan, thus avoiding conflicts between the tracks layer and the potential wells placed around each letter. That is, if the user is following a virtual fixture and he/she drags the cursor across a key, the elastic attraction effect around the key is not felt. When the user ends composing a word, he/she must press the white button present on the stylus of the Phantom to confirm the selection. It can happen that, especially for some commonly used characters, many completions are possible and, that, therefore, many virtual fixtures would be displayed by the SAPETS algorithm. We have noticed that this can be very disturbing since the attractive effects of the virtual fixtures tend to induce the user to take the pointer along the wrong track. Thus, we have limited the number of tracks that can be displayed to three. The displayed tracks are chosen on the basis of a statistical criterion: once a letter has been selected only the tracks corresponding to the completions of the three words more frequently introduced are displayed.

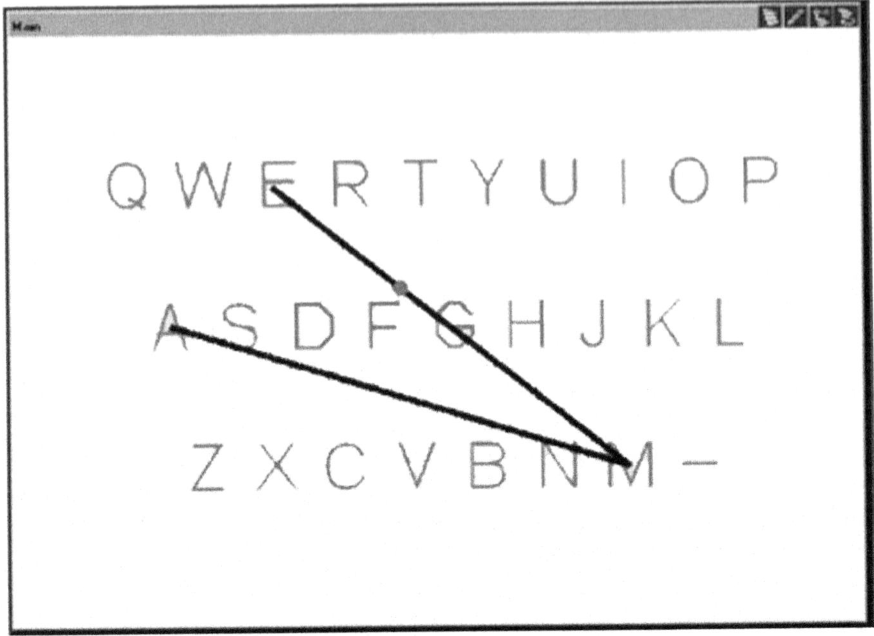

Figure 2 The virtual fixtures displayed using the SAPETS algorithm

To correct typing errors, the right button of a mouse was used. Since the aim of the experiments was to measure the time to type a string we assume that after an error all the typed characters are deleted and the person must restart the insertion process. The virtual keyboard together with some of the virtual fixtures that can be displayed is reported in Figure 2.

3. Experimental Results in case of single task

In the first set of experiments we compare the performances obtained by the users in the word insertion process in case the simple virtual keyboard is used with those obtained in case the virtual fixtures activated through the SAPETS algorithm described in Sec. 2 is used. The number of participants to the experiment was twenty four, all participants were right-handed with a mean age of twenty four years in a range from sixteen to forty nine. The user could move the pointer through the Phantom Omni and the virtual keyboard was displayed on a LCD monitor. The participants were divided into two groups, 12 persons each: in the first one, they operated on the first version of the virtual keyboard, namely without any word insertion facility, whereas in the second they could rely on the haptic layer of virtual fixtures described in Sec. 2. Each participant had to compose 24 words, 12 of which were classified as short ones, namely with less than 5 characters, and 12 as long ones, namely with more than 8 characters. All the words were in Italian, the mother tongue of all participants, since we wanted to avoid errors due to the scarce knowledge of foreign languages. In each group of words, sort ones and long ones, there were 9 common words, whose meaning was well known to all the participants, and 3 uncommon words, whose meaning was unknown by all the users. This choice has been done to simulate what can happen in a driving context, where sometimes the driver could have to insert as a GPS destination a string he/she has never used before (e.g. a small city where he/she has never been before). It was expected that participants would have carried out tasks more rapidly in the condition in which they could rely on virtual fixtures; the benefit deriving from the fixtures was expected to be detected with both short and long words.

Each participant was asked to follow this procedure:

1. read and understand the word
2. place the pointer in a predefined start position
3. when ready to start, push the blue button on the stylus;
4. for each letter of the word the procedure was:
    a. place the pointer in correspondence of the character on the keyboard
    b. push the white button on the stylus to confirm
    c. move to the next position
5. when the word is fully composed, press the blue button again.

After the execution of the test, the participants were asked to fill in a questionnaire presenting 5 questions about the usability of the system where the answers had to be selected on a 7-level Likert scale (Likert, 1932). Each participant was asked to do all actions naturally and without distraction. The movements of the participants' hands and the monitor were recorded during the execution of the experiment since we believe that the clip analysis may lead to the detection of features to improve in possible future developments. Execution times and errors were saved in log files. The use of fixtures (or not) and the word length (long or short) are the variables chosen for the performance analysis. For the group that was not using the virtual fixtures, the average times T_S^{NF} and T_l^{NF} for inserting a short word and a long word were respectively:

$$T_S^{NF} = 8.0823\ s. \qquad T_l^{NF} = 19.3752\ s. \qquad (1)$$

For the group that was using the virtual fixtures instead, the average times T_S^F and T_l^F for inserting the short words and the long words were respectively:

$$T_S^F = 8.0823\ s. \qquad T_l^F = 19.3752\ s. \qquad (2)$$

In Figure 3 we have reported a graphical representation of the average insertion time versus the word length. We can see that the virtual fixtures activated through the SAPETS algorithm introduce a beneficial effect by lowering the average insertion time. The benefits of the virtual fixtures become more relevant for long words. The activation of the virtual fixtures helps the user to rapidly move from one letter to the other on the virtual keyboard. The elastic potential well around the fixture helps the user to keep the right track so that he/she can rapidly move towards the desired letter.

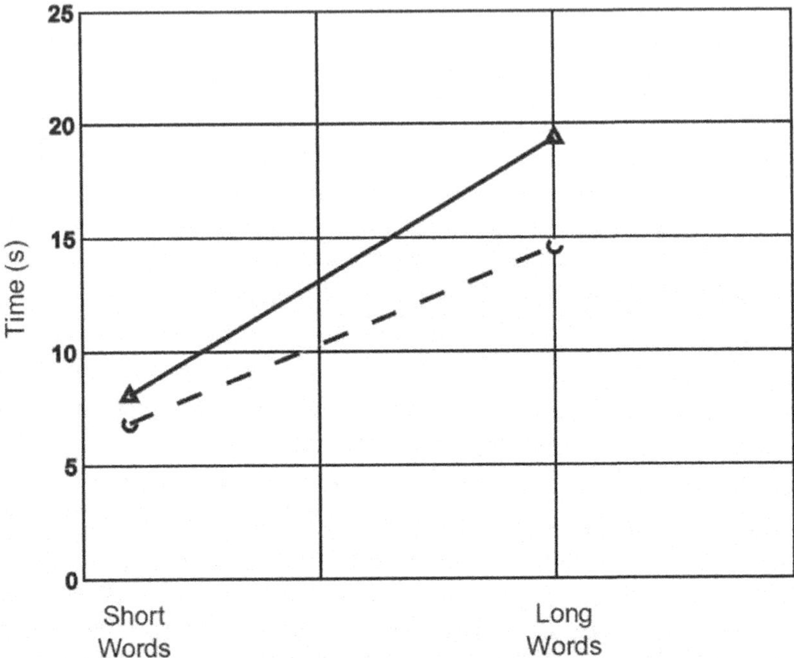

Figure 3 Average time for inserting the words with (dashed) and without (solid) fixtures The virtual fixtures displayed using the SAPETS algorithm

In order to analyze the errors committed by the users during the words insertion task, we have grouped the possible errors into 4 categories:

1. *Repeated key typing error:* a letter is selected more than once by pushing the button on the stylus.
2. *Near key error:* a letter near to the desired one is selected
3. *Accidental typing error:* a wrong letter is selected
4. *Distraction error:* a letter in the word is not inserted

In Table 1, the number of errors committed by the user during the experiment are reported

Table 1. Errors Analysis

Error Category	With Fixtures	Without Fixtures
Repeated key typing	29	9
Near key	1	4
Accidental typing	4	4
Distraction	4	10

The group of participants that use the virtual fixtures usually commits less errors. In particular, near key and distraction errors are significantly less when using fixtures. Thus, it seems that the presence of the virtual fixtures decreases the level of distraction and, consequently, the number of errors. The repeated key typing errors in the group that use fixtures is surprisingly higher than in the other group. Thus, it seems that the presence of the fixtures tends to make the user more unsure about the selection of a letter. Nevertheless, this kind of errors can be easily filtered via software. We aim at experimentally studying this phenomenon more in detail in the future. Once that the benefits of the haptic layer have been tested, it is necessary to evaluate the usability of the interface. In fact, if the users found the input device hard to use and if the system required too much attention for being used, it wouldn't be suitable for being used for the execution of a secondary task since it would tend to distract the user too much. In order to assess the usability of the interface, we have asked to each participant to evaluate, in a scale from 1 (absolutely no) to 7 (absolutely yes) the following statements:

1. generally the system is easy to use;
2. the system helps me to easily complete the assigned task;
3. the system has all the required functions and capacities
4. it is easy understanding when I commit an error;
5. the use of the system is intuitive.

The average evaluation of each statement is collected in Table 2.

Table 2. Interface Evaluation

Statement	With Fixtures	Without Fixtures
1	5.57	5.33
2	5.58	5.33
3	5.92	4.92
4	5.5	5.08
5	5.67	5.75

All the statements were given a high assessment; this means that the interface is perceived intuitive and easy to use. It is remarkable that, in case the virtual fixtures are enabled, the assessments of the statements increase or remain comparable with the corresponding ones in case no fixtures are used; this means that the introduction of the haptic word insertion facility is positively perceived in terms of the usability of the interface. In summary, we have experimentally proven that, in case the input task is the only task that the user has to complete, the virtual fixtures activated through the SAPETS algorithm lead to an improvement of the performance which seems to grow larger as long as the task becomes more complex (i.e. words become longer). Thus, the presence of the virtual fixtures is beneficial for the usability of the input device. Nevertheless, in the experiments illustrated in this section, the user can see the virtual keyboard and, therefore, he/she can rely both on the haptic and the visual information. What these results cannot tell is whether the above mentioned benefits would persist in case the input task becomes the secondary task and the visual demand for the primary task becomes quite high (e.g. in a driving context). This situation will be analyzed in detail in the next section.

3. Experimental Results in case of multiple tasks

The aim of the second experiment is to evaluate whether the virtual fixtures setup keeps on introducing benefits when the word insertion task becomes a secondary task, as it usually happens in a driving context (e.g. the insertion of a destination in a GPS while driving), and most of the visual attention of the user has to be drawn to a primary task. For this experiment, we have developed a graphical application, represented in Figure 4. It consists of one light grey sphere and four dark grey spheres. During the experiment, the light grey sphere is always visualized while the four dark grey spheres appear all together randomly for short periods of time.

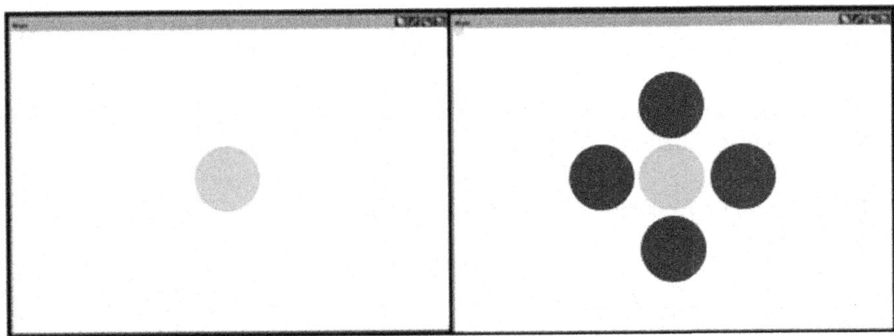

Figure 4 Snapshots of the graphical application that plays the role of the primary task

Each time that the dark grey spheres appear on the screen, the user has to push the space bar. The application allows to count both the number of times that the dark grey spheres appears on the screen and the number of times that the user presses the space bar in correspondence of their appearance. This application provides a simple but meaningful way for implementing a primary task which significantly captures the visual attention of the user. The difference between the number of times that the dark grey spheres have appeared and the number of times that the user has correspondingly pressed the space bar is an indicator of the distraction level of the user. The number of participants to the experiment was again twenty four, all participants were right-handed with a mean age of twenty six years in a range from twenty one to fifty eight. Each user was in front of two LCD monitors. The graphical application just described was displayed in one of the monitors and it played the role of the primary task. Each time that the dark grey spheres appeared on the screen, the user had to press the space bar. At the same time, each user was asked to compose 24 words (12 of which were classified as short ones, namely with less than 5 characters, and 12 as long ones, namely with more than 8 characters; all the words were in Italian and 3 words per each group were uncommon) using the virtual keyboard, displayed on the other LCD monitor, together with the Phantom omni as described in Sec. 2. The participants have been divided into two groups of 12 people. A group has used the virtual keyboard without the fixture system while the other one has exploited the virtual fixture facility. The secondary task is a source of distraction from the primary task. We have seen in Sec. 3 that the presence of the virtual fixtures helps the user to insert the words more rapidly. As a result of this experiment, we expected that the users exploiting virtual fixtures would have kept on inserting the words more rapidly than the other users. Furthermore, we expected that the presence of virtual fixtures would have decreased the distraction of the users from the primary task. The participant was asked to follow this procedure:

1. read and understand the first word
2. place the pointer in a predefined start position
3. when ready to start, push the blue button on the stylus; at the same time, the graphical application playing the role of the primary task starts
4. each time that the dark grey spheres appear, press the space bar
5. for each letter of the word the procedure was:
    a. place the pointer in correspondence of the character on the keyboard
    b. push the white button on the stylus to confirm
    c. move to the next position
6. when the word is fully composed, push the blue button again and start over with a new word

After the execution of the test, the participants filled in a questionnaire presenting 5 questions about the usability of the system where the answers had to be selected on a 7-level Likert scale. The participant was asked to do all actions naturally. The movements of the participants' hands and the monitors were recorded during the execution of the experiment. Execution times, the number of times that the user pressed the space bar and the number of times that the dark grey spheres had appeared in the graphical application were saved in log files. Firstly, we aim at assessing whether the use of fixtures keeps on introducing a benefit in the word insertion task also when a primary task is present. As in the previous experiment, the use of fixtures (or not) and the word length (long or short) are the variables chosen for the performance analysis. For the group that was not using the virtual fixtures, the average times T_S^{NF} and T_L^{NF} for inserting a short word and a long word were respectively:

$$T_S^{NF} = 7.7426\ s. \qquad T_L^{NF} = 24.1035\ s. \qquad (3)$$

For the group that was using the virtual fixtures instead, the average times T_S^F and T_L^F for inserting a short word and a long word were respectively:

$$T_S^F = 6.5598\ s. \qquad T_L^F = 19.8226\ s. \qquad (4)$$

In Figure 5 a graphical representation of the average times versus the length of the words has been reported.

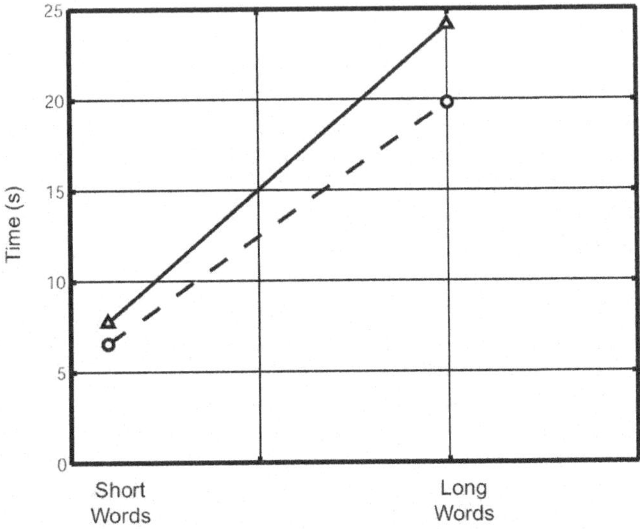

Figure 5 Average time for inserting the words with (dashed) and without (solid) fixtures

We can see that the virtual fixtures keep on introducing a beneficial effect in terms of velocity of insertion. Similarly to the results obtained in Sec. 3, the advantages of the virtual fixtures are more evident in case of the insertion of long words. The presence of the primary task increases the average insertion times with respect to the results obtained in Sec. 3 but the benefits introduced by the virtual fixtures are comparable with those obtained in Sec. 3. Thus, from the comparison of the experiments, it seems that the amount of benefit introduced by the virtual fixtures activated through the SAPETS algorithm is somehow independent of the presence of the primary task. In this second set of experiments we also want to evaluate the influence of the virtual fixtures layer in terms of performances of the primary task. The index used for evaluating the performances of each user on the primary task is the number of times he/she didn't detect the presence of the dark grey spheres because he/she was distracted by the word insertion process, namely by the secondary task. We have counted the total number of errors committed by each group, namely the total number of times that an appearance of the dark grey spheres hasn't been detected (i.e. the user didn't press the space bar) during the experiments involving members of the group. For the group that has not used the virtual fixtures the number of errors has been 42 while for the other group it has been 20. This result confirms our expectations: the presence of the virtual fixtures significantly improves performances also in the execution of the primary task. This improvement is mainly due to the fact that the fixtures allow the users to reach the desired letters over the keyboard exploiting mainly the haptic information leaving the user free to devote most of its visual attention to the primary task. In fact, once that the first letter of a word has been entered, the user has just to choose with a glimpse the direction taking the pointer to the next letter and then, thanks to the potential well around the fixture, he/she can let the pointer slide over the virtual track without any need to look at the keyboard. In this way, the time that the user spends at looking to the screen where the primary task is running dramatically increases as well as the number of committed error decreases. On the other hand, without the virtual fixtures, a significant portion of the visual attention of the user has to be devoted to the motion of the pointer over the keyboard and, consequently, he/she misses part of the dark grey spheres appearances. In other words, the presence of virtual fixtures allows to keep the visual attention of the user away from the secondary task which can be executed exploiting almost exclusively the haptic information. Also in this case, we found useful to evaluate the usability of the interface. We have asked to each participant to evaluate, in a scale from 1 (absolutely no) to 7 (absolutely yes) the same statements that have been reported in Sec. 3. The average evaluation of each statement is collected in Table 3.

Table 3. Errors Analysis

Statement	With Fixtures	Without Fixtures
1	5.25	4.46
2	5.92	4.61
3	6.5	4.5
4	5.76	5
5	6.15	4.5

Comparing these results with those reported in Table 2, we can see that the evaluations given by the group using the fixtures are, in average, higher. This means that the advantages of the fixtures is perceived by the users also in terms of usability of the interface. On the other hand, without fixtures, the distraction induced by the secondary task is perceived by the users in terms of a decrease of the usability; in fact the evaluations reported in the last column of Table 3 are lower than the corresponding ones in Table 2. In summary, we have experimentally proven that virtual fixtures activated through the SAPETS algorithm improve performance both in terms of velocity of insertion and in terms of errors committed in the primary task.

3. Experimental Results in case of multiple tasks

This paper presents a study on the use of virtual fixtures on input devices for secondary task. We have designed a virtual keyboard and we have proposed an activation algorithm, called SAPETS, for properly activating a set of virtual fixtures. We have conducted experiments that have proven that our algorithm makes the word insertion process faster. Furthermore, in case the input task is secondary and the primary task requires most of the visual attention of the user (e.g. a driving task), our algorithm decreases the distraction from the primary task. Future work aims at developing an input device endowed with virtual fixtures activated by the SAPETS algorithm to be embedded in the in-vehicle information system. Encouraged by the results of this paper, we believe that this input device will decrease the distraction from driving caused by the IVIS. We are building a prototype of a haptic input device that can enable virtual fixtures (see e.g. Wang, 2004) that can be easily integrated in a vehicle. It will be necessary to quantitatively evaluate the benefits of the virtual fixtures in a driving context. We will make experiments using a driving simulator that is being set up in our lab. Specifically, eye-tracking studies will be conducted in a simulated driving environment, in order to collect data regarding the actual visual demand imposed by the secondary task; such data are particularly relevant, since they directly impact on the eyes-off-the-road time, which is critical for drivers' safety.

References

Bengler, K., Herrler, M. e Künzner, H.: Usability Engineering bei der Entwicklung von iDrive (Usability Engineering accompanying the Development of iDrive). 3/2002, it - Information Technology, Vol. 44, p. 145. ISSN 1611-2776 (2002)

Bettini, A., et al.: Vision-assisted control for manipulation using virtual fixtures. IEEE Transactions on Robotics, Vol. 20, p. 953–966 (2004)

Donmez, B., et al.: A literature review of distraction mitigation strategies. SAVE-IT Project Deliverable. (2004)

Gibson, J.J.: The theory of affordances. R. Shaw e J. Bransford. Perceiving, acting, and knowing: Toward an ecological psychology. Hillsdale, NJ : Erlbaum, p. 67-82 (1977)

Likert, R.: A technique for the measurement of attitudes. Archives of Psychology, Vol. 140, 55, (1932)

Nolin, J.T., Stemniski, P.M., Okamura A.M.: Activation cues and force scaling methods for virtual fixtures. 11th Symposium on Haptic Interfaces for Virtual Environment and Teleoperator Systems, Chicago, IL, USA (2003)

Nowakowski, C., Utsui, Y. e Green, P.: Navigation system evaluation: The effects of driver workload and input devices on destination entry time and driving performance and their implications to the SAE recommended practice. The University of Michigan Transportation Research Institute.UMTRI-2000-20 (2000)

Payandeh, S., Stanisic, Z.: On application of virtual fixtures as an aid for telemanipulation and training. 2002. Proceedings of the Symposium on Haptic Interfaces for Virtual Environment and Teleoperator Systems (2002)

Rosenberg, L.: Virtual fixtures: Perceptual tools for telerobotic manipulation. Proceedings of the Virtual Reality Annual International Symposium (1993)

Secchi C., Stramigioli, S., Fantuzzi, C.: Intrinsically passive force scaling in haptic interfaces. Proceedings of IEEE/RSJ International Conference on Intelligent Robots and Systems. Beijing, China (2006)

Young, K., Regan, M. e Hammer, M.: Driver distraction: a review of the literature. Monash University Accidents Research Centre. Report 206. (2003)

Wang, D., et al.: Haptic overlay device for flat panel touch displays. Proceedings of the Symposium on Haptic Interfaces for Virtual Environment and Teleoperator Systems. Chicago, Illinois, USA (2004)

How *'learnable'* are CASE tools in diverse user communities?

Brenda Scholtz[1] and Janet Wesson[2]

Department of Computer Science and Information Systems (CS & IS), Nelson Mandela Metropolitan University, P O Box 77000, Port Elizabeth, 6061, South Africa

[1] brenda.scholtz@nmmu.ac.za
[2] janet.wesson@nmmu.ac.za

Abstract. The use of Computer Aided Software Engineering (CASE) tools for teaching object-oriented systems analysis and design (OOSAD) and the Unified Modelling Language (UML) has many potential benefits, but there are several problems associated with the usability and learnability of these tools. This paper describes a study undertaken to determine if computing students from a linguistically and technologically diverse community experience problems with learning to use a CASE tool, and to determine if there is a relationship between two user characteristics of the students and the learnability of CASE tools.

Keywords: CASE tool, usability evaluation, learnability, UML, language diversity

1 Introduction

The Unified Modelling Language (UML) has been made an OMG (Object Management Group) standard (Frosch-Wilke, 2003) and helps analysts specify, visualise and document models of software systems, and thus improve the chances of success of the project (Kemerer, 1992; Brewer and Lorenz, 2003). The use of Computer Aided Software Engineering (CASE) tools for teaching UML has a number of potential benefits but there are several problems associated with the usability and learnability of these tools (Lending and Chervany, 1998; Booch et al., 1999; Post and Kagan, 2000; Burton and Bruhn, 2004). Learnability is one of

Please use the following format when citing this chapter:

Scholtz, B. and Wesson, J., 2008, in IFIP International Federation for Information Processing, Volume 272; *Human-Computer Interaction Symposium*; Peter Forbrig, Fabio Paternò, Annelise Mark Pejtersen; (Boston: Springer), pp. 83–97.

the most important attributes of usability and refers to the capability of the system to enable the user to learn to use the application (Nielsen, 1993).

Some guidelines are available relating to the usability and learnability of UML and CASE tools (Jarzabek and Huang, 1998; Booch et al., 1999; Johnson and Wilkinson, 2003), however there is a lack of guidance regarding what is necessary to make a CASE tool 'learnable' and how to evaluate the learnability attribute of usability (Phillips et al., 1998). By performing CASE tool learnability evaluations, knowledge of the factors that influence the rate of learning can be determined, which can lead to improved approaches to teaching of CASE tools.

The South African university education system serves students from a wide range of backgrounds (Rauchas et al., 2006). There is enormous language diversity, with the country having 11 recognised official languages. The student community at a South African university can therefore be classified as linguistically diverse (Koch, 2002; Greyling and Calitz, 2003). Research has shown that the non-English speaking students are disadvantaged (Koch, 2002). These student communities have different frequency of computer user profiles and are thus technologically diverse. The home language and frequency of computer use are thus two important user characteristics for South African university students, and could affect the rate at which they learn to use a CASE tool.

The structure of the rest of the paper is as follows. Section 2 gives an overview of the concepts of usability and learnability, and explains Senapathi's framework for CASE tool learnability, which is used throughout this paper. Section 3 discusses the research design used and the results of the study are provided in Section 4. The paper ends with a discussion in Section 5 and concluding remarks in Section 6.

2 Background

2.1 Usability and learnability

Different classification schemes for quantifying and assessing usability have been proposed (Gould and Lewis, 1985; Nielsen, 1993; Barnum, 2002; Dix et al., 2004; Seffah et al., 2006) and they all specify learnability as one of the key attributes of usability. The ISO/IEC 9126-1 definition of usability is concerned with the attributes of the software system that make it understandable, learnable, easy to use and attractive (ISO, 2001). In order to determine how easy it is to learn to use a CASE tool for students in diverse user communities, a framework for evaluating OO CASE tool learnability was required. The first framework investigated was one proposed by Phillips *et al.* for evaluating the usability of CASE tools (Phillips et al., 1998).

The second framework reviewed was proposed by Senapathi and included only the learnability attributes of usability (Senapathi, 2005). Prior to the Phillips and Senapathi studies, methods and frameworks for evaluating CASE tools were mainly suitable for commercial environments (Mosley, 1992).

Since the focus of this study was on learnability, Senapathi's framework was selected as the most appropriate framework to use. An overview of this framework and the results of Senapathi's study is discussed in the next section.

2.2 Senapathi's framework for evaluating CASE tool learnability

A framework used to evaluate the learnability of CASE tools was designed by Senapathi based on the ISO 9241 definition of usability (Senapathi, 2005). This framework is illustrated in Figure 1. The framework proposes that the users' *Satisfaction* ratings of the *Learnability attributes* of a CASE tool are dependent on certain *Context of use* factors. These context of use or contextual factors include the tasks, the learning environment, the user characteristics and the CASE tool used.

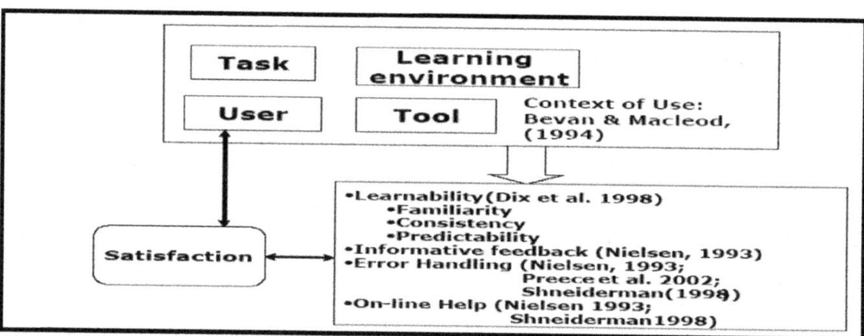

Figure 1 A framework for the evaluation of CASE tool learnability (Senapathi, 2005)

In an educational context, all the activities and assessments that require the use of a CASE tool are recognised as tasks (Senapathi, 2005). Senapathi's framework recognises the significance of the learning environment in the learnability evaluation of CASE tools in educational environments (Senapathi, 2005). The background and context in which the course is delivered should be studied and analysed. This includes the learning methodologies, teaching methods and resources used. CASE tools should be learnable in a short timeframe and support a wide range of different learner characteristics due to the time constraints of students (Senapathi, 2005). Senapathi's study explored the effects of five user characteristics on learnability. These were Gender; General level of computer experience; Previous experience with CASE tools; Number of hours spent per week with CASE tools; and Attitude and motivation.

The Tool aspect of the framework relates to the complexity of the selected CASE tool and how this complexity affects learnability (Senapathi, 2005). Senpathi's study used Rational Rose as the selected CASE tool. Satisfaction was used as a direct measure of evaluating whether or not a CASE tool is learnable in a given context (Senapathi, 2005).

Senapathi proposed six attributes that should be used to measure learnability: Familiarity, Consistency, Predictability, Informative Feedback, Error Handling and On-Line Help. Familiarity is the extent to which a user's knowledge and experience in other real-world or computer-based domains can be applied when interacting with a new system (Dix et al., 2004). Consistency is the likeness in behaviour arising from similar situations or similar task objectives (Dix et al., 2004). Predictability is the ability of the system to allow the user to anticipate the natural progression of each task. The presence of appropriate and relevant feedback, specifically in the context of handling error messages, is considered to have a significant effect on the learnability and understandability of a system (Jankowski, 1995; Norman, 1999). The quality of feedback provided by the CASE tool is an important learnability measure (Phillips et al., 1998). Human Computer Interaction (HCI) research highlights the importance of the provision of help facilities in software for improving usability and learnability (Nielsen, 1993; Post and Kagan, 2000; Senapathi, 2005; Seffah et al., 2006).

2.3 Results and limitations of Senapathi's study

The results of Senapathi's study showed that three of the five user characteristics tested had significant effects on learnability. These were Computer experience; Previous experience with CASE tools; and Attitude and motivation. The other two user characteristics showed no significant results. Senapathi's study also revealed that all groups rated the Error handling features of Rational Rose significantly lower than the other features, regardless of their user characteristics, specifically with regard to Consistency and Feedback (Senapathi, 2005).

Senapathi's study has several limitations. The first limitation is the fact that it was only tested at the University of Auckland, New Zealand and may not be suited for educational institutions with linguistically and technologically diverse communities where English is not the home language of the majority of the student population. The tool part of the framework was not verified as the study only included an evaluation of one CASE tool, namely Rational Rose.

The first section of the learnability questionnaire used by Senapathi included closed-ended questions that were grouped according to the learnability attributes in the framework. Ease of Learning, was not listed as a learnability attribute in Senapathi's original framework, but the first group of questions in Senapathi's learnability questionnaire were grouped under the heading, *Ease of Learning*. Ease of learning refers to the novice user's experience on the initial part of the learning curve while trying to learn a new system (Nielsen, 1993).

Although On-line help was listed as a learnability attribute in Senapathi's framework, no closed-ended questions relating to On-line help were included in the learnability questionnaire.

The next section discusses the research design for an experiment undertaken to validate Senapathi's framework at a South African university.

3 Research design

The primary research question of this study was to determine what CASE tool learnability problems are experienced by students in a diverse community. The second research question was to determine if there is a relationship between CASE tool learnability and the user's context of use. In order to answer these questions and to verify Senapathi's framework in a diverse user community, an experiment was performed in 2006. The participants of this experiment were students registered for an OOSAD course at a South African university, and were representative of the student population of that found in a typical university in South Africa.

The experiment included a usability evaluation of two CASE tools using the framework proposed by Senapathi. Two additional user characteristics were identified, namely home language and frequency of computer use (Section 1), which were not included in Senapathi's study. In order to verify the Tool section of the framework two CASE tools were selected, namely Microsoft Visio and IBM's Rational Software Modeller (a more recent version of Rational Rose). Microsoft Visio was selected as the least complex of the two CASE tools since it has fewer features than Rational Software Modeller (Scholtz, 2003).

3.1 Hypotheses

The null hypothesis "*H_0: No relationship exists between the learnability of a CASE tool and the context of use*" was formulated for examination and tested for significance at the 95% significance level ($\alpha = .05$). It was refined to produce the sub-hypotheses in Table 1.

Table 1 Research Hypotheses

Number		Hypotheses
$H_{0.1}$		No learnability problems exist in either of the selected CASE tools.
$H_{0.2}$		No relationship exists between the learnability of a CASE tool and the type of tool used.
$H_{0.3}$		No relationship exists between the learnability of a CASE tool and the user characteristics.
	$H_{0.3.1}$	No relationship exists between the learnability of a CASE tool and the user's home language.
	$H_{0.3.2}$	No relationship exists between the learnability of a CASE tool and the user's frequency of computer use.

3.2 Evaluation instruments

The post-test learnability questionnaire used in Senapathi's study was amended by adding a second questionnaire for the second CASE tool, which had an additional section at the end where the participant had to select his/her CASE tool preference as well as the reason for the preference. An attitude and motivation questionnaire was completed by all participants on completion of both tasks, where they were required to rate their attitude towards the use of a CASE tool in an OOSAD module. This questionnaire was the same as the one used in Senapathi's study.

3.3 Participant and task selection

Participants recruited for this experiment were students at a South African university enrolled for the Information Systems 2.1 module. Both groups were taught the same material on UML prior to the tasks and had no previous experience of CASE tools prior to starting the experiment. For Task A, the participants were required to draw a use case diagram, and for Task B, a class diagram. Sixty-two students agreed to participate in the study and stratified sampling was used to divide these students into two equivalent groups. Prior to commencing the first task, a background questionnaire was completed by each participant.

Most of the existing questionnaires, scales and tests that have been developed in highly industrialised countries of Europe and North America cannot be applied to the South African context without some modifications as the user communities in these countries are not as linguistically or technologically diverse as those in South Africa (Mouton, 2001).

Users who have different home languages and different frequency of computer use profiles may have different preferences for user interfaces and this could have an effect on the learnability of a CASE tool (Ford and Gelderblom, 2003; Shneiderman and Plaisant, 2005). Two additional user characteristics, namely, home language and frequency of computer use, were therefore added to the background questionnaire.

The profile of the participants according to home language and frequency of computer use is shown in Figure 2. Only 28% of the participants had English as their home language, and almost half (46%) had a Low frequency profile (less than 10 hours per week computer use). The students were randomly assigned to each group, while making sure that the proportion of both gender and academic performance per group was representative of the actual student population. Each participant was required to complete both tasks in both CASE tools.

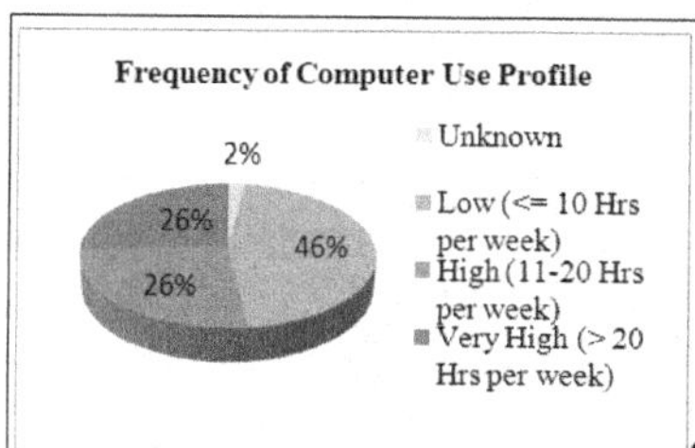

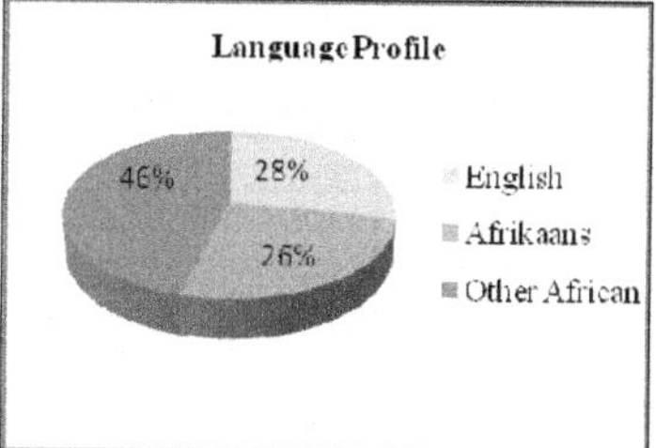

Figure 2 Profile of participants

The experimental design method used was a counterbalanced design and the same design was used for both tasks. The tasks were identical for both groups, except for the order of the CASE tools used. The one group used Visio first and the other group used Rational first. After each task, the participants completed a post-test learnability questionnaire for each CASE tool.

4 Research results

Post-test questionnaires from only 46 of the original 62 participants could be included in the data analysis, as the remaining 16 participants did not attend both tasks.

4.1 Quantitative results

The first section of the post-test learnability questionnaire included 15 closed-ended items relating to the learnability attributes proposed in the framework and participants were asked to rate each item on a Likert scale of 1 to 5, where 1="predominantly disagree" and 5="predominantly agree". Initial statistics on the responses from the first section of the learnability questionnaire revealed that Visio scored significantly higher than Rational for both tasks for all attributes except Familiarity. Figure 3 illustrates the mean satisfaction ratings for each learnability attribute for both tasks.

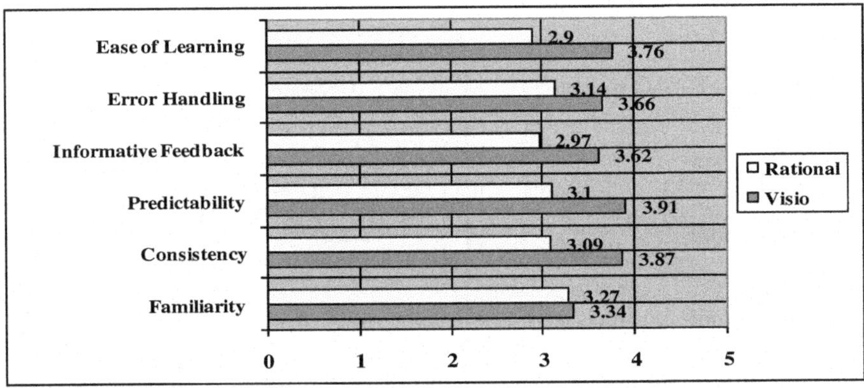

Figure 3 Learnability results for both tasks (5-point Likert scale)

The highest rated attribute for Rational for both tasks was Familiarity (with a mean rating of 3.27), whereas for Visio it was Predictability (with a mean rating of 3.91). The lowest rated attributes were Ease of learning for Rational (with a mean rating of 2.9) and Familiarity for Visio (with a mean rating of 3.34). From this we can deduce that the participants rated the two tools differently in terms of the six attributes of the CASE tools measured, and that the attributes that were rated best and worst for the two CASE tools were not the same. These results were obtained irrespective of whether the tool was used first or second. A repeated measures ANOVA test was conducted on the three factors and the overall mean ratings for the different learnability attributes. The first factor was the type of CASE tool used. The results of these tests showed that there is a definite relationship between the CASE tool used and learnability as the repeated measure for the tool used was significant for all of the attributes except Familiarity for both Task A and Task B. For both tasks, for all of the learnability attributes (except Familiarity), there was therefore an interaction in the mean ratings of both CASE tools from performing the task in the first tool and then in the second tool. Visio was rated higher than Rational irrespective of the order in which the tools were used.

The results of the ANOVA tests also revealed that a relationship exists between both of the user characteristics and learnability; namely home language and frequency of use. There was a significant interaction between the mean rating for Predictability for the Afrikaans language group and the African language group ($p = .030$). The mean rating for the Afrikaans language groups for Predictability dropped from the first CASE tool to the second CASE tool in Task A, whereas the African language group's mean rating increased. The mean rating for the English language group remained fairly constant from one tool to the other, which could be due to the fact that the language used in the user interface of both CASE tools is English. Frequency of computer use was significant for only one attribute, namely Error Handling, for both tasks. The interaction occurred between the Low (<= 10 hours per week) and the High (11-20 hours) frequency groups, for both tasks.

Visio was selected the preferred tool for Ease of Learning for both tasks by an overwhelming majority for Task A (78%) and for Task B (72%). The most frequent reason for selecting Visio as the preferred tool for Ease of Learning, was Simplicity. The participants were also asked to rate their attitude towards the use of a CASE tool in the OOSAD module. The mean rating per question is shown in Table 2.

Table 2 Attitude and motivation towards use of CASE tools in teaching OOSAD (5-point Likert scale)

Question	n	Mean	SD
The use of a CASE tool for this course was a good idea.	41	4.10	0.80
The CASE tool made my work more interesting	41	3.71	0.81
The use of a CASE tool enabled me to complete my tasks more quickly.	41	3.68	0.85
The use of a CASE tool helped me to understand the underlying concepts better.	41	3.73	0.78
Correct understanding and use of the CASE tool helped me to perform better in the UML section of the course and assignments.	41	3.80	0.78
Overall Attitude and Motivation	**41**	**3.80**	**0.68**

The results showed a favourable response towards using a CASE tool for teaching OOSAD and that using a CASE tool helped to understand the underlying UML concepts.

4.2 Qualitative results

The qualitative data collected from the responses to the open-ended questions were coded and structured into categories which were then matched to the corresponding learnability attribute. Frequency counts for each attribute were calculated in order to determine the most common attributes. The most frequent reason for selecting Visio as the preferred CASE tool for Ease of Learning, was Familiarity. One of the participants stated that Visio was "*More familiar*" and another that it was *"Easier to learn and use since it has the same interface as Word and Excel"*.

Participants were also required to indicate what they did not like about each CASE tool and any problems encountered. Several learnability problems with both CASE tools were identified. For both tools the attribute with the highest frequency count for negative comments was Recoverability. One of the problems encountered by the participants was Visio's handling of erroneous associations between UML shapes. Visio highlights in red any part of the diagram which is in violation of UML rules, and displays the related error message in the error window below the drawing pane. Many of the students did not like this approach as they did not know why the association line was highlighted in red, due to the fact that the related error message was often not seen. One participant stated "*The red lines indicate errors but these weren't obvious what they meant.*" Some of the students found Rational's approach to error prevention and handling very confusing as the CASE tool does not allow an erroneous association between shapes to be made. It does not display any message explaining why the association cannot be drawn, which led to frustration and confusion on the part of the student. One participant stated that "*When you make a mistake the program does not allow you to perform the action at all. The program is confusing to use*" Other participants made similar comments. The attribute that had the second most negative comments was Customisability for Visio. Most of the responses relating to this attribute were with regard to the inflexibility of the editing of the communication line names. The attribute with the most negative responses for Rational was Simplicity. It can be deduced from this that many of the participants found Rational to be too complex. This confirms the selection of this CASE tool as the more complex tool (Section 3).

4.3 Hypotheses testing

Several learnability problems were encountered by participants in both CASE tools, Microsoft Visio and Rational Software Modeller (Section 4.2), therefore $H_{0.1}$ "*No learnability problems exist in either of the selected CASE tools*" can be rejected. As can be seen from the initial quantitative results reported (Section 4.1), the learnability ratings of the two CASE tools was significantly different for five of the six learnability attributes measured.

This was confirmed by the more detailed repeated measures ANOVA tests showing that the type of CASE tool used has an effect on learnability (Section 4.1). One can thus conclude that $H_{0.2}$ "*No relationship exists between the learnability of a CASE tool and the type of tool used*" should also be rejected.

The statistics showed that there was a relationship between the home language of participants and some of the learnability attributes (Section 4.1), therefore the hypothesis $H_{0.3.1}$, "*No relationship exists between the learnability of a CASE tool and the user's home language*" was rejected. Initial results from the repeated ANOVA tests revealed that frequency of computer use was significant for one attribute, namely Error Handling, for both tasks (Section 4.1). The hypothesis $H_{0.3.2}$ '*No relationship exists between the learnability of a CASE tool and the user's frequency of computer use*' was therefore rejected. Since both $H_{0.3.1}$ and $H_{0.3.2}$ were rejected, $H_{0.3}$ can be rejected. Since $H_{0.1}$, $H_{0.2}$ and $H_{0.3}$ can all be rejected, this implies that H_0 "*No relationship exists between the learnability of a CASE tool and the context of use*" can be rejected. Table 4 summarises the research hypotheses and includes an indication whether the hypotheses can be rejected or not.

Table 4 Summary of hypotheses and results

Number		Hypotheses	Result
$H_{0.1}$		No learnability problems exist in either of the two CASE tools.	Rejected
$H_{0.2}$		No relationship exists between CASE tool learnability and the type of tool used.	Rejected
$H_{0.3}$		No relationship exists between CASE tool learnability and the user characteristics.	Rejected
	$H_{0.3.1}$	No relationship exists between CASE tool learnability and the user's home language.	Rejected
	$H_{0.3.2}$	No relationship exists between CASE tool learnability and the user's frequency of computer use.	Rejected

5 Discussion

One of the primary research questions of this study was to determine what problems students in a diverse community in South Africa experience with learning to use a CASE tool. The results of the experiment performed at a university in South Africa show that the majority of problems identified related to error prevention, error handling and feedback.

Microsoft Visio was clearly the preferred CASE tool for all the attributes of learnability. The attributes of learnability that were rated the highest by the students for their preferred CASE tool were Predictability, Consistency and Ease of learning. Designing CASE tools that are predictable, consistent and easier to learn can thus improve the rate of learning of CASE tools as well as indirectly improve the understanding of UML. The results of this study identified problems with the way in which both CASE tools tried to constrain UML errors, although the approaches used were very different. Visio's problems related to the way in which the erroneous UML connection was highlighted in red. Participants were not aware of the reason for this as the error message did not use language which was easy to understand and the message was displayed at the bottom of the screen. Rational uses error prevention and does not allow the participant to perform any erroneous UML connections. This frustrated the participants as they did not know why they were not allowed to perform the action.

One way of overcoming the problems with both of these approaches would be to use a confirmation error message when preventing an erroneous UML connection (Nielsen, 1994). The differences in users' reactions to the two approaches, as well as any proposed improvements in error handling, need further research before any conclusions can be made. A relationship was found to exist between the home language and frequency of computer use of the participants and their ratings of the two CASE tools, regardless of the order in which the tools were used. Further investigation with larger sample sizes is required in order to examine this relationship in more detail. Further research should be undertaken by extending the study to other universities in South Africa, which would provide a wider range of home language and frequency of computer use profiles. In order to be able to extend the results to commercial environments, similar evaluations could be undertaken at commercial organisations in South Africa.

6 Conclusions

This paper has investigated the learnability of CASE tools for education in a diverse student community at a university in South Africa. It discussed an experiment conducted at a South African university to address the limitations of Senapathi's framework in a South African context. The results of this research show that significant differences in learnability were found between the two CASE tools investigated. Relationships between two user characteristics and learnability were also identified, namely home language and frequency of computer use. The results also identified that simplicity is a key factor for learnability and that existing error handling techniques are problematic. These results can be used to improve the learnability of CASE tools and assist educators in selecting suitable CASE tools for diverse user communities.

References

Barnum, Carol (2002): *Usability Testing and Research.* Pearson. 0-205-31519-4.

Booch, G, Rambaugh, J and Jacobson, I (1999): *The Unified Modelling Language User Guide.* Addison-Wesley.

Brewer, J and Lorenz, L (2003): Using UML and Agile Development Methodologies to Teach Object-Oriented Analysis & Design Tools and Techniques. In *Proceedings of CITC4'03*, Lafayette, Indiana, USA:54-57. ACM. October 16-18, 2003.

Burton, P and Bruhn, R (2004): Using UML to facilitate the teaching of object-oriented systems analysis and design. *Journal of Computing Sciences in Colleges*, 19(3):1937-4771. January 2004.

Dix, Allan, Finlay, Janet, Abowd, Gregory D and Beale, Russell (2004): *Human-Computer Interaction.* Third Edn, Prentice Hall.

Ford, Gabrielle and Gelderblom, Helene (2003): The Effects of Culture on Performance Achieved through the use of Human Computer Interaction.

In *Proceedings of SAICSIT 2003*, Gauteng, South Africa. South African Institute for Computer Scientists and Information Technologists.

Frosch-Wilke (2003): Using UML in Software Requirements Analysis - Experiences from Practical Student Project Work. In *Proceedings of InSITE - Where Parallels Intersect*. Informing Science. June.

Gould, J.D. and Lewis, C. (1985): Designing for usability: key principles and what designers think. *Communications of the ACM*, 28(3):300-311.

Greyling, Jean and Calitz, Andre (2003): The Development of a Computerised Multimedia Tutorial System for a Diverse Student Population. In *Proceedings of 2nd International Conference on Computer Graphics, virtual reality, visualisation and interaction in Africa AFRIGRAPH '03*, Cape Town, South Africa:109-116. ACM.

ISO (2001): *ISO/IEC 9126-1: Software Product Evaluation* [online]. Available at http://www.iso.org/iso/iso_catalogue/catalogue_tc/catalogue_detail.htm?csnumber=22749. [Accessed on July 2006].

Jankowski, DJ (1995): Case feedback in support of learning a systems development methodology. *Journal of Information Systems Education*, 7(3):88-90.

Jarzabek, S and Huang, R (1998): The Case for User-Centered CASE tools. *Communications of the ACM*, 41(8):93-99. August 1998.

Johnson, H.A and Wilkinson, L (2003): CASE tools in object-oriented analysis and design. In *Proceedings of Eastern Conference: Consortium for Computing Services in Colleges*.

Kemerer, Chris (1992): How the Learning Curve Affects CASE tool adoption. *IEEE Software*, 0740-0759/92/0500/0023. May 1992.

Koch, E (2002): Language Testing at UPE. In *Proceedings of Language Testing Colloquium*, UPE.

Lending, D and Chervany, N (1998): CASE tools: Understanding the Reasons for Non-Use. *ACM SIGCPR conference on computer personnel research SIGCPR'98*, 19(2). April 1998.

Mosley, V (1992): How to Assess Tools Efficiently and Quantitatively. *IEEE Software*, 9(3):29-32. May 1992.

Mouton, J (2001): *How to succeed in your Master's & Doctoral studies*. Pretoria, Van Schaik Publishers.

Nielsen, J (1993): *Usability engineering*. Academic Press.

Nielsen, Jakob (1994): *Ten Usability Heuristics* [online]. Available at www.useit.com. [Accessed on February 2007].

Norman, D.A (1999): Affordance, conventions and design. In *Proceedings of SIGCHI Interactions*, 6(3):38-42. May 1999.

Phillips, C, Mehandjiska, D, Griffin, D, Choi, M.D and Page, D (1998): The usability component of a framework for the evaluation of OO CASE tools. In *Proceedings of Software Engineering, Education and Practice*, Dunedin:134-141. IEEE. 26 - 29 January 1998.

Post, G and Kagan, A (2000): OO-CASE tools: an evaluation of Rose. *Information and Software Technology*, 42:383-388. 15 April 2000.

Rauchas, S, Rosman, B, Konidaris, G and Sanders, I (2006): Language Performance at High School and Success in First Year Computer Science. In *Proceedings of SIGCSE '06*, Houston, Texas, USA. ACM.

Scholtz, Brenda (2003): *A Comparative Analysis of CASE Tools*. Honours Treatise. Department of CS & IS, University of Port Elizabeth. Port Elizabeth.

Seffah, Ahmed, Donyaee, Mohammed, Kline, Rex and Padda, Harkirat K (2006): Usability measurement and metrics:A consolidated model. *Software Quality Journal*, 14:159-178.

Senapathi, M (2005): A Framework for the Evaluation of CASE tool learnability in Educational Environments. *Journal of Information Technology Education*, 4.

Shneiderman, Ben and Plaisant, Catherine (2005): *Designing the User Interface*. Fourth Edn. 0-321-19786-0.

A Prospect of Websites Evaluation Tools Based on Event Logs

Vagner Figuerêdo de Santana[1] **, and M. Cecilia C. Baranauskas**[2]

[1] Institute of Computing, UNICAMP, Brazil, v069306@dac.unicamp.br
[2] Institute of Computing, UNICAMP, Brazil, cecilia@ic.unicamp.br

Abstract: The variety of websites evaluation tools based on event logs is growing, but some of their limitations are already visible (e.g., the need for task model, plug-in dependency, use of simulated tasks, separation of accessibility from usability, etc). Some of these characteristics result in coupled systems and make the configuration and use of these tools more expensive. This work aims to show the main features and weaknesses of these tools. One expects that the discussion and the requirements pointed out in this paper could help developers of evaluation tools so they could reuse consolidated ideas and avoid identified weaknesses.

Keywords: website evaluation tools, log based evaluation, event loggers

1. Introduction

In the last years many researchers have been working on automatic and remote usability evaluation of user interfaces focusing on tests with more users without resulting in costly evaluations (Ivory, 2001). The total or partial use of automatic usability evaluation methods can reduce time and costs involved in the development of a Web application, as it liberates specialists from repetitive tasks as manual log analysis (Paganelli, 2002). While researchers have focused on usability issues, less has been done specifically to increase the accessibility level of the applications. In addition, the integration of Accessibility and Usability (A&U) concepts is not usual in user interface (UI) evaluation tools.

One of the aspects that make interesting the approach of automatic evaluation is the remote evaluation during real use, and involving users in their regular environments. In addition, remote analysis of A&U make unnecessary the presence of participants in the test places, which could be a barrier. Moreover, tests in controlled environments are artificial and may influence the results (Rubin, 1994). Thus, the UI evaluation in real situations points to the use of event logs, since it makes possible to capture data while users work on their regular tasks, in their usual environments (Guzdial, 1993).

Please use the following format when citing this chapter:

de Santana, V.F. and Baranauskas, M.C.C., 2008, in IFIP International Federation for Information Processing, Volume 272; *Human-Computer Interaction Symposium*; Peter Forbrig, Fabio Paternò, Annelise Mark Pejtersen; (Boston: Springer), pp. 99–104.

Interface events have a duration ranging from 10 milliseconds to one second (Hilbert, 2000), showing that the number of events that occur during few minutes simple tasks can be huge. Thus, due to the typical amount of interface events, the use of automatic evaluation tools is usually necessary so that information extracted from the events reach a level of abstraction that is useful to specialists (Hilbert, 2000). This paper identifies solutions, limitations and gaps of website evaluation tools based on event logs; it is organized as follows: section 2 presents ideas well succeeded and the gaps identified; section 3 discusses the results of this work; section 4 presents conclusions and points to new directions.

2. Event logs and websites evaluation: identified solutions, limitations, and gaps

UI events are natural results of using windows based interfaces and their components (e.g., key strokes, mouse clicks) (Hilbert, 2000). Since it's possible to record these events and they indicate the user behavior during the interface usage, they represent an important source of information regarding usability. Event logs produce results as frequency of use of certain functions, places where users spend more time and the sequence they complete tasks (Woo, 2004).

According to Hong et al. (Hong, 2001), the goal of evaluation software based on event logs is to use a capture method that is easy to apply to any website, and that is compatible with various operating systems and web browsers. This section summarizes the main issues present in the following tools: WET (Etgen, 1999), WebRemUSINE (Paganelli, 2002), and WebQuilt (Hong, 2001). Also, we combine some ideas with other works related to the evaluation of websites as, for example, NAUTICUS (Correani, 2006), a tool that integrates A&U and points to a promising direction for new tools. Our analysis is organized according to eight aspects: the configuration needed to use the tools, the capture mechanisms, the ways logs are stored and transmitted, the dependence on users' actions, the changes needed at the UI, scalability, levels of abstraction, and integration of A&U. We present the solutions, limitations we found, and gaps to be investigated.

1) Regarding the configuration of the environment, i.e. the set of tasks needed for a tool to be used to evaluate a website, we can point out:

Solutions – Events capture on the client-side at WET depends only on the native script languages of the newer browsers (e.g., JavaScript), avoiding the need of plug-ins installation. Other solutions that make possible to capture across different websites without requiring any change on evaluated Web pages are the use of a proxy logger to capture server-side data (Hong, 2001), or the use of specific browser to capture client-side data (Claypool, 2001).

Limitations – Require maintainers of the evaluated website to change their Web pages so that the data capture starts, as in (Paganelli, 2002; Etgen, 1999); or that files must be hosted on the server of the evaluated websites (Etgen, 1999). Other important points are the dependency of task models (Paganelli, 2002) and dependency of specific software installed, as in (Paganelli, 2002; Claypool, 2001).

Gap - Events capture at client-side across different websites without requiring any change in the evaluated web pages or using specific browsers.

2) Regarding server-side vs. client-side capture, server-side capture concerns the use of logs generated only on the server, while client-side capture refers to data obtained through the recording of information available only on the client's device.

Solutions – Use of events triggered at client-side interface, because there are more detailed data sources of how the user uses a Web site (Paganelli, 2002; Etgen, 1999). Also, the use of a proxy logger to record all communication between the client's browser and the visited Web pages, without requiring any change in the evaluated pages or any plug-in installed in the client's device (Hong, 2001).

Limitations – The limitations related to the data capture deals with the storage of data too, because the amount of data obtained through events on the client-side is representative and the space available to store them in client's device is restricted, as in (Etgen, 1999). On the other hand, when there is space available on the client-side, the problem arises from the time spent on the log transmission at the end of a test session, as can occur in (Paganelli, 2002). An alternative would be to use only server-side data, but it can lead to less accurate results (Claypool, 2001).

Gap – The combination of server-side data and the client-side data, since server data may indicate events not present on the client-side (e.g., response times of accessing a link). On the other hand, client-side data have more detailed information about the user route in a Web page.

3) Storing and transmitting logs refers to strategies used to record event logs, including the location, data storage capacity, and the transfer methods to a server.

Solutions – Use of a proxy logger that captures the data about client requests and server responses (Hong, 2001). Thus, the data storage capacity depends on the tool server capacity and it's not necessary to send data to another element. To avoid the limit of available space in cookies, some components (e.g., Java applets) can be used to store and transmit client-side events (Paganelli, 2002).

Limitations – When recording data on a participant machine, the tool becomes dependent on the available space in that device, as in (Etgen, 1999). Moreover, the transmission of such data at the end of a test session, as in (Paganelli, 2002; Etgen, 1999), may require too much bandwidth connection, and thus interfere with the use of the interface. Another identified limitation deals with the use of software components that are plug-ins dependent (e.g., Java applets), as in (Paganelli, 2002). This can be an obstacle, since the user may not have knowledge to install the plug-in or doesn't have a broadband connection to download it quickly.

Gap – The use of cyclic transmission of event logs from client-side to the server, avoiding the limit of available space in cookies and the time required to transfer captured data in batch at the end of a test session.

4) By dependence on user actions we mean what is necessary from the participants for the tool to start the data capture (e.g., acceptance of the use).

Solution – The user accesses the first URL from the main page of the tool and then no further action is required until the end of the test session (Hong, 2001).

Limitation – Interfere the common use of the interface, requiring that the user select a task from a list or select options to start and stop the capture every test session. This may call the user's attention or even not allow that the data between these actions be captured, as in (Paganelli, 2002; Etgen, 1999).

Gap – An alternative to requiring a user action at the beginning of each test session is to make the acceptance of a test to be valid for several sessions, until the user chooses to stop the data capture.

5) Regarding the impacts that the evaluated UI suffers when a tool is in use, the key metric used is how the presentation of the tool interfere with the use of the website under evaluation.

Solutions – Display options to indicate the beginning and the end of capture (Etgen, 1999) or using a Web page, at the beginning of each session (Hong, 2001).

Limitations – The interface changes concern the reduction of the useful space for the UI to display any component of the evaluation tool, as occurs in the presentation of the list of tasks in (Paganelli, 2002). Another issue is the change of URLs used in the pages processed by the tool, as in (Hong, 2001).

Gap – Ensure that the tools' controls are always visible, don't compete with other UI elements, and don't reduce the useful area of the Web page.

6) Scalability refers to how the evaluation tools deal with many evaluation sessions or long test sessions.

Solutions – The use of specific software (e.g., web browser, applet Java) that can access participant's file system to record logs and then avoid data storage capacity of cookies, as in (Paganelli, 2002). Or use of a proxy logger to keep track of user requested data (Hong, 2001), depending only on the tool's server.

Limitation – The biggest barrier found was the space available for storage of logs in cookies, as it is not possible to record all usage, it forces the events captured to be a reduced sub set of existing events, as in (Etgen, 1999).

Gaps – Avoid limits of cookies without making the tool plug-ins dependent and avoid to transfer logged data only at the end the tests sessions.

7) As concerns the strategies used to achieve higher levels of abstraction from logs of events (e.g., implicit interest, use of grammars to relate events).

Solutions – Using task models and grammars that enable events of Web pages to be represented as higher levels actions (Paganelli, 2002) and conversion of low level events into sequences of actions (Hong, 2001).

Limitation – Dependency of specific task models and/or grammars to reach higher levels of abstraction, as in (Paganelli, 2002).

Gap – The use of methods that do not depend on task models to obtain higher level information. A possible alternative to these models is the sequence characterization based on the Markov Chains (Hilbert, 2000).

8) Finally, the integration of A&U is necessary, because if developers evaluate them separately, then some problems may appear regarding different priorities for guidelines interfering in the way the target audience uses an interface.

Solutions – Development and use of criteria covering aspects of A&U and evaluation of structure and content of pages (Correani, 2006).

Limitation – The identified limitation found in NAUTICUS is the dependence of a dictionary of terms commonly used in bad structured pages.

Gap – The integration of A&U was not found in websites evaluation tools based on logs, thus becoming the main gap of the studied tools.

3. Discussion

In this study we could identify characteristics that can strengthen evaluation tools, making them more robust, less costly, and easier to use and reuse. The integration of A&U has been identified as the main gap in the analyzed tools. We propose requirements that a logs based websites evaluation tool should have (Table 1). The requirements are organized in the Semiotic Ladder (SL), an artifact of Organizational Semiotics (Stamper, 1993) which provides a framework for analysis of information systems under six different layers. The SL structure allows the clarification of concepts and the analysis of information, considering from the Information Technology (IT) Platform (i.e., physical layer, empirical layer, and syntactic layer) to the Human information Functions (i.e., semantic layer, pragmatic layer, and social layer).

Table 1. Requirements for websites evaluation tools based instantiated in the SL

	Human Information Functions
Social	Focus on the integration of A&U for the target audience of the evaluated website. Enable remote testing during real use of the evaluated website. Interfere with the Web page as minimum as possible.
Pragmatic	Provide controls representing the status of the tool and user context during the test session. The tool should use two actions: one to start the capture, which stay valid for future sessions, another to interrupt the capture, which may occur at any time.
Semantic	Provide high levels of abstraction without depending on specific task models, grammars, or events.
	Information Technology Platform
Syntactic	Use all available data (e.g., client-side events and server-side logs) in order to obtain correlations between them. The combination of the available data in different components can reveal information impossible to obtain independently.
Empirical	Prevent that processing or transmitting logs interfere with the use of evaluated interface. The tool should implement safe and effective techniques without impacting on the website usage.
Physical	Do not depend on resources or specific configuration of the participants devices (e.g., disk space, bandwidth, etc). The evaluation tool should include mechanisms to achieve their goals in different configurations of hardware and software.

Analysis shows that the tools studied meet, mainly, the requirements associated with layers related to IT Platform (i.e., physical, empirical, and syntactic layers). The requirements of the Human Information Functions (i.e., semantic, pragmatic, and social layers) are hardly addressed. This suggests that new tools must take them into account to achieve goals related to these layers.

4. Conclusion

This paper presented a study of websites evaluation tools based on event logs, as well as different solutions to problems encountered at the event capture, log transmission, etc. The main results of this work are related to the identification of requirements and promising investigation subjects. Thus, developers of new websites evaluation tools can avoid the weaknesses and address some unexplored aspects such as the integration of A&U, for example. Based on the characteristics present in websites evaluation tools based on event logs, a set of requirements was proposed and have to be addressed to avoid the limitations discussed in this work.

The next steps of this investigation involves the development of a website evaluation tool based on logs that combine the solutions in order to fill as much as possible the identified gaps, taking into account the requirements listed.

References

Claypool, M., Le, P., Wased, M., Brown, D.: Implicit interest indicators. In: IUI '01: Proceedings of the 6th Int. conference on intelligent user interfaces, New York, NY, USA, ACM (2001) 33–40.

Correani, F., Leporini, B., Paternò, F.: Automatic inspection-based support for obtaining usable web sites for vision-impaired users. Universal Access in the Information Society **5**(1) (2006).

Etgen, M., Cantor, J.: What does getting wet (web event-logging tool) mean for web usability? In: Proceedings of 5th Conference on Human Factors & the Web. (1999).

Guzdial, M.: Deriving software usage patterns from log files. Technical report, Georgia Institute of Technology (1993).

Hilbert, D.M., Redmiles, D.F.: Extracting usability information from user interface events. ACM Comput. Surv. Vol. 32 (4) (2000) 384–421.

Hong, I., J., Heer, J., Waterson, S., Landay, A., J.: Webquilt: A proxy-based approach to remote web usability testing. ACM Transactions on Information Systems 19 (3) (2001) 263–285.

Ivory, M.Y., Hearst, M.A.: The state of the art in automating usability evaluation of user interfaces. ACM Comput. Surv. Vol. 33 (4) (2001) 470–516.

Paganelli, L., Paternò, F.: Intelligent analysis of user interactions with web applications. In: IUI '02: Proceedings of the 7th int. conf. on intelligent user interfaces, ACM (2002) 111–118.

Rubin, J.: Handbook Of Usability Testing: How to plan, design, and conduct effective tests. 1st edn. John Wiley & Sons Inc (1994).

Stamper, R.: A semiotic theory of information and information systems / applied semiotics. In: Invited papers for the ICL/University of Newcastle Seminar on "Information", (1993).

Woo, D., Mori, J.: Accessibility: A tool for usability evaluation. In Masoodian, M., Jones, S., Rogers, B., eds.: APCHI. Volume 3101 of LNCS, Springer (2004) 531–539.

Habbo Hotel – Academic Studies in Mixed Feelings

Raija Halonen[1], **and Eeva Leinonen**[2]

[1] University of Oulu, Department of Information Processing Science, Finland, Raija.Halonen@oulu.fi
[2] University of Oulu, Department of Information Processing Science, Finland, Eeva.Leinonen@oulu.fi

Abstract: "The first time was not so painful they claim!" This is an example of comments expressed by university students who were compelled to use Habbo Hotel in their studies. In this article we analyse the usefulness of an internet game called Habbo Hotel as a collaborative platform when the students carry out their studies on digital media. Today, digital media has evolved to concern several dimensions of everyday life. We want to understand how digital media bends to act as social media in teaching and learning. Social media is characterised by participation, openness, conversation, community and connectedness. These concepts built the core also in our lessons where human-computer interaction was emphasised.

Keywords: education, teaching with games, collaborative, virtual interaction

1. Introduction

"Only by putting my shoulder to the wheel I succeeded to make a Habbo personality." (A male student in his blog November 20, 2007). This quotation shows the difficulties the adult students perceived in the beginning when they participated in a new way of carrying out an academic course. This paper discusses the impressions and challenges we met in our course in the Department of Information Processing Science. We wanted to study how human-computer interaction with virtual communities persuades students' learning or is it only a burden. The first attitude was skeptic and the students took the chosen media with a grain.

According to Mayfield (2007) social media can be defined as a group of new kinds of online media which share the following characteristics: participation,

Please use the following format when citing this chapter:

Halonen, R. and Leinonen, E., 2008, in IFIP International Federation for Information Processing, Volume 272; *Human-Computer Interaction Symposium*; Peter Forbrig, Fabio Paternò, Annelise Mark Pejtersen; (Boston: Springer), pp. 105–117.

openness, conversation, community and connectedness. Due to convenience, we chose Habbo Hotel (Habbo, 2007) as our platform in the course where we taught digital media. "Habbo is a virtual world where you can meet and make friends." Habbo Hotel is a virtual community owned and operated by Sulake Corporation (Wikipedia, 2007a). According to Wellman & Gulia (1999) virtual communities are studied as communities but we want to add to this with our research on virtual community as a persuasive technology of human-computer interaction to support teaching and learning.

We used the means of qualitative case study in our research despite research that grounds on rich qualitative data predict challenges (Eisenhardt & Graebner, 2007). The challenges were increased in our study due to the chosen platform.

This article is organised as follows: The next section brings forward literature review on virtual communities, problem-based-learning and www-based games. After that, the research methods are presented and the case is briefly described. That is followed by empirical illustrations of the case. After that, discussion and results are presented ending with suggestions for future research.

2. Literature Review

Communality as a concept can be defined by communal limits, perceived safety, togetherness and a joint language (MacMillan & Chavis, 1986). Social media is something more: Participation, openness, conversation, community and connectedness are the keywords that define modern social media Mayfield (2007) continued. Participation refers to contributions and feedback from everyone who is interested. Openness lets people to give feedback, to participate and to share information. Furthermore, conversation is realised by two-way interaction between those involved. Community refers to the possibility to form communities quickly and easily around common interests. Finally, connectedness means that social media is dependent on the possibility to combine different kinds of media in one place. (Mayfield, 2007)

On the other hand, the term "community" is sometimes used to refer to intensive social groups whereas on the other occasions it refers to a group of people who hardly know each other. Every now and then the term "community" is used as a concept that means a geographic place. (Etzioni & Etzioni, 1999)

Etzioni and Etzioni (1999) compared face-to-face communities and virtual communities with the help of six factors: access to the communication, verification in communities, interactive broadcasting, reassembling, meeting mechanisms and memories. They claim that it is an illusion to assume that real-world communications are much more effective than communications in virtual environments. Etzioni and Etzioni continue that combining real-world and virtual communications could allow the special strengths of each system to make up for weaknesses of the other.

The characteristics and processes related to utilitised virtual community were studied by Blanchard and Markus (2004). Their study expresses that the perceived sense of community is characterised by social processes of exchanging support, creating identities and making identifications and by the production of trust. These processes seem to be similar to those processes that contribute to the formation of sense of community without virtual reality. Blanchard and Markus continue that a sense of virtual community does not even take form without the recited processes.

However, according to Li (2004) it is significant to notice the differences between virtual communities and virtual teams or groups. Li defines that virtual teams are formed to solve specific problems or tasks. Contrary to that, virtual communities focus on relationships in real life especially when people have no intended reasons to remain in the communities. Furthermore, virtual communities may exist for a long time while virtual groups or teams tend to disappear after the specific task or function is completed. Preece (2000) adds that virtual communities necessitate people who interact when they strive for their goals; common meaning, strategies and information systems that support social interaction and enable a feeling of togetherness.

Li (2004) argues that a plausible definition of virtual communities is still missing and that studies on virtual studies are still in their initial stage. Therefore, more research should be carried out especially by the means of case studies and empirical studies, Li concludes.

In all, one can conclude that human-computer-aided communities are not yet studied in full. Our article adds to the research with an empirical study that consists of a case study on using virtual communities as enablers and platform when teaching graduate students.

3. Research Methods and the Case

In this chapter the research methods are expressed. Furthermore, due to the role of the learning technique in the context, also it is essentially described. Finally, the case is briefly explained.

This study is qualitative case study (Stake, 2000) in nature. Eisenhardt (1989) delineates case study as a research strategy that focuses on understanding the dynamics present within single settings. Eisenhardt also supposes that case study research has important strengths like novelty, testability and empirical validity which arise from the close linkage with empirical evidence. Theory building from case studies is an increasingly popular and relevant research strategy that forms the basis of an excessive large number of significant studies (Eisenhardt & Graebner, 2007).

In our study, we observed the meetings and collected written material written by the participants. Furthermore, we documented virtual sessions that were held during the observations. We also collected emails and other notes that were written related to the course. To analyse the empirical material, we used discourse analysis (Potter, 1999) that supports naturally occurring talk and discussions in their venue. The discussions are emphasised in the research material due to the nature of the case even if it was demanding to get the talk recorded.

Problem-based learning (PBL) is an instructional strategy of "active learning" that is often used in higher education and students are encouraged to take responsibility for their group and organize and direct the learning process with support from a tutor or instructor (Wikipedia, 2007b). As a learning technique, PBL can be characterised by three essential assertions: 1) understanding is in our interactions with the environment; 2) cognitive conflict or puzzlement is the stimulus for learning and determines the organisation and nature of what is learned; 3) knowledge evolves through social negotiation and through the evaluation of the viability of individual understanding (Savery & Duffy, 2001). PBL aims to help students gain content knowledge, strengthen problem-solving skills, engage in learning and develop a professional identity (Shumow, 2001).

The case consisted of carrying out an undergraduate course that was a part of digital media studies in the Department of Information Processing Science. The course was compulsory or optional depending on the orientation of the student. The course was to be performed by active participation or by a literature exam. Already beforehand, the active way was advertised in the course web site:

"The course is performed by active participation both in contact lessons and in www-based working in a way that will be announced in the course. In addition to group and individual tasks, there will be an exam that will be explained in detail later in the course. On top of this description, the course can be performed by a literature exam."

Habbo Hotel is a virtual hotel, where "teenagers can hang out and chat" (Habbo, 2007). As of July 2007, over 82 million avatars have been created worldwide and there are 6 million unique visitors to the virtual hotels around the world every month, and 75,000 avatars being created every day (Wikipedia, 2007).

The educational background with its problem-based learning (PBL) was introduced in the first lesson and the chosen platform (Habbo Hotel) was described to the students. Every student was told to make an own Habbo personality (avatar) i.e. register him or her self (Figure 1).

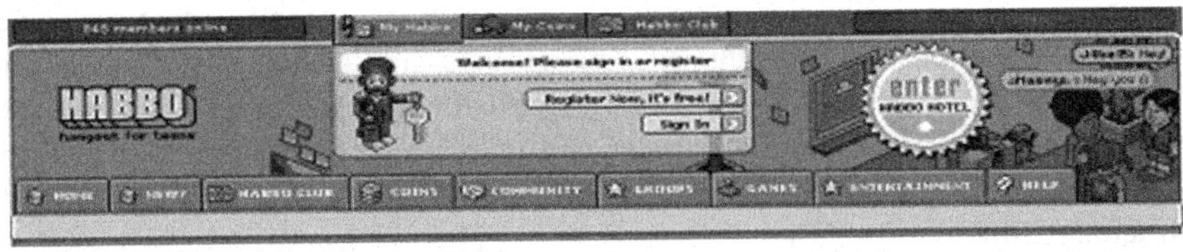

Figure 1 Entrance of Habbo Hotel.

Despite the richness of colour in the welcoming window (Figure 1) the user finds it easy to proceed. In our case, the notice of "hangout for teans" was experienced foolish.

4. Empirical Illustrations and Analysis

In this chapter the empirical findings are described in detail. The informative description enables the reader to get a full understanding on the case and challenges that were met there. The empirical research material is collected during an undergraduate course that was carried out by "active performance".

In our case, the activity necessitated blog-writing, PBL technique in learning and using Habbo Hotel (Habbo, 2007) in group meetings. This procedure was introduced to the students in the first so-called compulsory lecture. After having heard the procedure, several students left the lecture hall. A pronounced resistance to the procedure was almost touchable.

The participation necessitated signing in the course but the signing was enabled also after the first lecture. Because the core topic was digital media and digital communities with their diversified culture the teachers had chosen Habbo Hotel as a learning platform in the course. The feature of Habbo Hotel being recommended for children above ten years old gave the teachers challenges to introduce the media to the students who were well over 20 years of age. The students commented: *"Making a Habbo personality sounds chiefly surrealistic …"*. In our case there were almost 100 students who participated in the course. That amount led to 13 groups with five to six students. The groups were formed at the end of the first lecture and only some groups consisted of students who were familiar to each other from the very beginning.

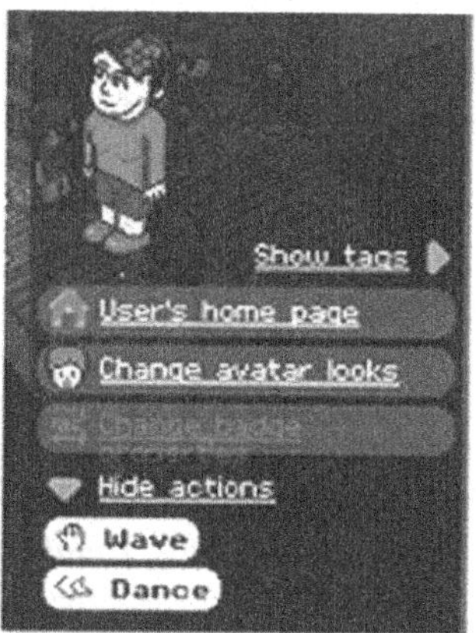

Figure 2 A Habbo avatar that can wave or dance.

The first homework was to make a blog for the group and to make an avatar (Habbo personality) for each student. Figure 2 shows an avatar that can be made to wave her hand or dance. When preparing the course, the teachers already had created their own Habbo personalities and named them with a prefix Ed to distinguish from students' names. The students were directed to name their avatars so that they could be identified. Despite the wish, that desire did not come true in the course. Creating Habbo personalities was more difficult to some students because it appeared that operating systems acted differently in relation to the needed Shockwave. The students wrote their blogs: *"On top of all this, Habbo is difficult to reach by almost everyone except Windows users ..."* Fortunately this problem was solved with the help of Habbo: "So how do you get into Habbo if you have an Intel based Mac? Simple! Follow these steps: […] " (Habbo, 2007). This difficulty in human-computer interaction connected with technology increased the perceived resistance to adopt Habbo as the learning platform.

Besides the students, also the teachers perceived problems when building the learning platform. A course blog was created in the very beginning and it was announced to act as the primary information source in relation to all course notices and communication. To relieve the burden of the teachers in following the group blogs the group blogs were to be included as RSS entries in the course blog. However, only nine RSS entries were enabled and another way to manage the group blogs was to be found due to the great amount of the groups. Finally, the group blogs were added in a blogroll of the course blog. The group blogs were listed in the course blog but it was not possible to get the updated information automatically fed i.e. as RSS feed. Therefore, the teachers had to visit every group blog to see if there were any new annotations in the blogs and that increased their work a lot.

The first homework was checked in the next lecture. Additional students joined the course and they were ordered to form a group. The teachers recorded: *"The change is remarkable. This lecture hall was dark with students; there were not enough chairs, even. Some questions were expressed but we were prepared to more of them. I am surprised at this reception. We also had to admit that the teachers cannot visit every session because there are so many groups. Therefore we must flexibly apply PBL here."*

The students were encouraged to write their group-based blogs about their experiences with Habbo Hotel. The notes were diversified. Some students were open-minded and they wrote positive annotations: *"... the course implementation appears interesting and at least different ...". "The first ambiance is a little mixed but for once there is something new in the studies, this blogging and habboing."* Others students were attacking with their comments: *"To my mind, the start was chaotic due to the new implementation." "My weekend is spoilt. Thank you Wordpress, thank you Internet!"* Some students seemed to be satisfied and thanked the change: *"Extraordinary implementation is always a plus, whatever course it concerns."*

The actual learning topic was introduced next week after the students had found a group for themselves and registered in Habbo Hotel. The topic "The special characteristics of virtual communities" was announced in the course blog and it included an article "Net Surfers Don't ride alone: Virtual Communities as Communities" (Wellman & Gulia, 1999). The students were to study the topic by the means of PBL. Three questions were given to them to be answered during their learning:

How is community realised in the internet?

Which kind of communities are there in the internet?

What is combining these communities to "virtual communities"?

Following the procedure of PBL, the students were to have a brainstorm in the Habbo room. The yellow spots in the figure 3 represent the tags that were used in the brainstorm.

Figure 3 Group in a session.

The discussion in the group was carried out by balloons that drifted up as new balloons were written (Figure 3). In case of active discussion the speed was greater and it appeared difficult to follow the discussion. According to the chosen learning methodology PBL, the discussion should have been recorded by a secretary and a chairman should have taken care of the order in the group. A third role was called an observer. Observers should have observed the social dynamics in the session. Instead of being permanent, the roles had to be circulated among the students. One role lasted only one session. This requirement was set to give an equal work load for the students and also to enable an active participation in discussions. Writing the discussions down appeared such a burden that the secretaries were not able to discuss actively in the same session.

The students were very censorious with the discussion platform. They complained the limited space in the balloons that prevented long sentences in addresses: *"For some reason the maximum length of the messages is limited in Habbo so that it is impossible to write sensible sentences in one message."* Another topic for complains was the lacking possibility to log message history: *"By the way, the hotel expressed its best character because the message history is still missing in the discussions in Habbo."*

It also happened that the software broke down every now and then: *"The occasional falls of Habbo troubled working. Furthermore, it was irritating that the comments went to the heaven of bits."* In order to relieve the interaction, some groups used additional tools in their group meetings: *"The meeting starts at 6 pm. We'll gather both in IRC and Habbo Hotel."*

Besides the students, also the teachers used additional tool during the group meetings. The teachers decided to use Skype's [18] chat that enabled them to interact so that the students were not able to see the comments. Skype also enabled the private discussions to be saved for later use.

An interesting feature of the chosen media was the possibility to make the Habbo avatar "speak" in loud, whispering or normal voice. That feature was necessary when the avatars were located in public rooms (e.g. bar or lounge in the hotel) but it was not needed when the students gathered in their classrooms. In case the students did not realise to "shout", the addresses were not properly visible in the balloons. This situation was realized when the avatars were situated too far from each other (Figure 4).

Figure 4 An avatar (Tonyyh) is speaking too quietly.

Some of the virtual sessions looked so comical out that the teachers felt it difficult to participate without laughing. The funny outlooks also tempted to use informal words and phrases when talking with the students in Habbo Hotel. The environment simply forced to act friendly and cosy. The teachers wrote to each other: *"It will be difficult to meet face-to-face these students after this session!"*

Unfortunately it was not possible to observe what happened in the Habbo room without being virtually present. Therefore, the teachers had to "step in" the same Habbo room with the students when they wanted to supervise the students.. In this sense, it was also impossible to know how the teachers' intervention influenced the interaction. On top of being present in the room, the teachers had to be located near enough in order to "hear" the discussion. The teachers regretted: *"It is a pity that we cannot see what they do there when we are not present."* It also happened that the free discussion did not proceed e.g. when the students agreed on roles in the meeting. The teachers wrote to each other in Skype: *"Half an hour is passed and they are still discussing roles!"* When it looked like the students were not able to proceed in their assignment the teachers occasionally left the room. The teachers continued: *"They are not listening to what we tell them." "This isn't progressing at all."*

The students took the platform with mixed feelings. The teachers participated in a session where one student tried to push others to change to IRC (Internet Relay Chat). He was very persistent in his arguments. The teachers chatted: *"Javelin is pushing them to IRC but the others do not want to."*

Nonetheless, the teachers would have liked to see how the groups worked and how they managed to use Habbo in their interaction and learning. They needed to know who was active and how the given problem was solved. Occasionally the teachers observed: *"They are pretty quiet."* From the viewpoint of the course on digital media studies, the chosen platform acted both as a target and a tool.

Figure 5 Avatars with differing outlooks.

It appeared that the outlook of the avatars was perceived very important because the window was relatively small and the creatures too small to be properly identified. Figure 5 displays a session where there were two groups at the same time in the hotel room. Without the exceptional outlooks it could have been impossible to identify the students or distinguish them from each other.

The groups got their assignments done and they reported their findings in their blogs. Some of them wanted to compare different platforms: *"The only issue that Habbo Hotel offers compared to IRC is the graphic user interface. That may help the younger generation to make a better feeling of community but it is not so essential to the older people. Especially if more important features are left out because of that."*

One group had deliberated dating in internet: "The most important feature in the profile of dating in internet is the image they give of the person's personality."

The groups also found some benefits in virtual communities: "Giving feedback is easier in virtual communities because the tool acts as a buffer between the giver and the receiver. In some virtual community it is problematic that the giver may hide behind anonymity and she or he may give feedback that would not be given as real self." An interesting notice was about participation: "In virtual community it is much easier not to participate in the community activities than in real-worlds communities".

All these quotations from the students' blogs illustrates that the students were able to use Habbo Hotel when they discussed the given problems. They also were able to conclude their occasionally fragmental sentences that they wrote in Habbo Hotel. On the other hand this new way to realize the course was led to participation and collaboration of the students. Also Barker (1999) and Xing and Spencer (2008) reported advantages for students' active approaches in virtual learning context.

5. Discussion

In this article we have intimately explored digital media in the form of Habbo Hotel (Habbo, 2007) as a platform of human-computer interaction in teaching graduate students. In addition to contribute teaching, also learning is an important issue to be concerned. We used qualitative research methods in our research and the main analysis was carried out by the help of discourse analysis. The research material consisted mainly of notes made of private discussions between two teachers, blogs written by the students and sessions in Habbo Hotel. The research material offered a rich empirical material that enabled research based on empirical study, thus responding to the appeal of Li (2004). Next we summarise the baseline:

First, the ground for the study was fruitful and sensible. The research material was collected in a course that was a part of digital media studies in the Department of Information Processing Science. This starting point would support the progress and motivation and also prefigure the output of the research.

Second, the digital media was a game that was designed for children above ten years of age. This feature was a challenge but it was recognised in the very beginning, before the course had its start. The feature also predicted easy human-computer interaction.

Third, the participating students had a choice to perform the course in another way. Thus participating in the "active way" was voluntary.

Fourth, the chosen media (Habbo Hotel) was not designed for meetings, but that was known beforehand. In addition, the platform was not designed for teaching purposes either. Despite that, the interaction in the social community could consist of learning by PBL.

At the beginning of the course, the reaction of the students was astounded and staggering. There were more students present than was planned beforehand. Also the perceived problems with technology at the beginning caused irritation. After the first reaction, the atmosphere changed more towards expectant. The blogs contained notes that exposed both negative and positive opinions.

Due to the unexpected amount of students, the organising of the course appeared to be difficult and laborious. In the course there were several small virtual communities: each group was told to form a virtual community and they should have met in Habbo Hotel one group at a time. The teachers were called to participate every virtual meeting. However, it happened that occasionally two groups gathered simultaneously to the Habbo room. The students also contemplated in their sessions if the whole course formed one community. Thus, the environment (Habbo Hotel) really supported their thinking and learning.

Because of the limited usability of Habbo Hotel, both the students and the teachers felt it important to use additional media during the sessions. The main additional tool used by the students was IRC and the teachers used Skype, respectively. The number of carried discussions varied between the groups but at the end every group had been able to ponder the given problems and they had written constructive comments on the topic.

From the viewpoint of the teachers, the interaction between the teachers and the students seemed to be easier in the Habbo room. The sentences had to be kept short and also the words seemed to contain more information than in real-life. A bare question mark was needed if somebody wanted to spell "I beg your pardon". It also seemed that the tone in the sentences was friendlier than when interacting face-to-face.

Despite the experienced friendly and constructive performance in the Habbo room, there were students who insisted to change the whole forum (e.g. Javelin, see above). However, these students were minority and the group work was mainly carried out in the settled environment, even if supported with other media.

Furthermore, we would like to emphasise the informal nature of the forum. The environment with the funny-looking creatures supported open-minded interaction. On top of the outlook of the avatars, the avatars could also be made dancing. They could "dance" all the session through because the "dance" was an on-off-feature of the avatar.

In all, one can conclude that the experiment teaching in Habbo Hotel was a positive one and both the students and the teachers learnt a lot. Grounding on the thorough reports on the meetings in Habbo Hotel and the written blogs on the given topic we figure that Habbo Hotel may contribute teaching and learning with its limited human-computer interface. Despite Habbo Hotel was designed for teenagers, it proved to be useful also in adulthood.

In our case we had too many students in the course and the teachers were not able to participate as they had planned. However, they could monitor the progress by reading the blogs. In this sense, the blogs were complementing the interaction between the students and the teachers. The blogs also completed the interface that was experienced insufficient in certain moments.

To conclude, we recommend future research on using such platforms in teaching. We suggest that other platforms will be used and experiments collected. It would be worth studying such environments that also offer easy possibilities to record discussions. We also emphasise that avatars should be kept in the forthcoming research palette because that gives possibilities to make comparisons between these studies. Furthermore, we also claim that avatars as such contribute free-floating and unprejudiced discussion.

References

Baker, P.: Using Intranets to Support Teaching and Learning. Innovations in Education and Teaching International, vol 36 iss 1, pp. 3-10 (1999).

Blanchard, A.L., Markus, L.M.: The Experienced "Sense" of a Virtual Community: Characteristics and Processes. Database for Advances in Information Systems 35;1, 65-79 (2004).

Eisenhardt, K.M.: Building Theories from Case Study Research. The Academy of Management Review 14;4, 532-550 (1989).

Eisenhardt, K.M., Graebner, M.E.: Theory building from cases: opportunities and challenges. Academy of Management Journal 50;1, 25-32 (2007).

Etzioni, A., Etzioni, O.: Face-to-Face and Computer-Mediated Communities, A Comparative Analysis. The Information Society, 15, 241-248 (1999).

Habbo:, Habbo Hotel. (2007) Available: http://www.habbo.com/. Accessed 14 April 2008.

Li, H.: Virtual Community Studies: A Literature Review, Synthesis and Research Agenda. Proceedings of the Americas Conference on Information Systems, New York, New York, August (2004).

MacMillan, D.W., Chavis, D.M.: Sense of Community: A Definition and Theory. Journal of Community Psychology 14;1, 6-23 (1986).

Mayfield, A.: What is social media, an ebook from spannerworks. USA: Spannerworks. (2007). Available: http://www.spannerworks.com/fileadmin/uploads/eBooks/What_is_Social_Media.pdf. Accessed 14 April 2008.

Potter, J.; Discourse Analysis as a Way of Analysing Naturally Occurring Talk. In: Silverman, D. (ed.) Qualitative Research. Theory, Method and Practice, pp. 144-160. SAGE Publication, London (1999).

Preece, J. Online Communities – Designing Usability, Supporting Sociability. John Wiley & Sons, Chichester (2000).

Savery, J.R., Duffy, T.M.: Problem Based Learning: An instructional model and its constructivist framework. CRTL Technical Report No. 16-01 (2001).

Shumow, L.: Problem-Based Learning in an Undergraduate Educational Psychology Course. In: Levin, B. (ed.): Energizing Teacher Education and Professional Development with Problem-Based Learning. Association for Supervision & Curriculum Development Alexandria, VA, USA, 24-44 (2001).

Skype. Available: http://www.skype.com. Accessed 14 April 2008.

Stake, R.E.: Case studies. In: Denzin, N.K., Lincoln, Y.S. (eds.) Handbook of Qualitative Research, pp. 435-454, Sage Publications Inc. Thousand Oaks (2000).

Wikipedia, (2007a). Available: http://en.wikipedia.org/wiki/Habbo_Hotel. Accessed 14 April 2008.

Wikipedia, (2007b). Available: http://en.wikipedia.org/wiki/Problem-based_learning. Accessed 14 April 2008.

Wellman, B., Gulia, M.: Net surfers don't ride alone: Virtual communities as communities. In: Kollock, P., Smith, M. (eds.). Communities and Cyperspace Routledge New York (1999).

Xing, M., Spencer, K.: Reducing Cultural Barriers via Internet Courses. Innovations in Education and Teaching International, vol 45 iss 2, pp. 169-181 (2008).

IMPROVING ACCESSIBILITY TO GOVERNMENTAL FORMS[1]

Norbert Kuhn, Stefan Richter, Michael Schmidt, Andreas Truar

Institut für Softwaresysteme in Wirtschaft, Umwelt und Verwaltung
Fachhochschule Trier, Umwelt-Campus Birkenfeld, Postfach 1380, D-55761 Birkenfeld, Germany,

{n.kuhn|s.richter|s.naumann|m.schmidt}@umwelt-campus.de

Abstract: Although many governmental institutions have provided their costumers with access to electronic government documents there is still a lack of accessibility for handicapped citizens. In this paper we present an approach to improve access to governmental forms for handicapped citizens, in particular for people with visual impairments, elderly people, illiterates or immigrants. We describe a system where we combine scanned images of paper-based forms containing textual information and a text-to-speech synthesis to an audio-visual document representation. We exploit standard document formats based on XML and web service technology to ensure independence from software and hardware platforms.

Keywords: Multimodal User Interfaces, Accessibility, Forms Reading for E-Government

1. Introduction

In recent years much effort has been spent in Human Computer Interfaces to improve access for handicapped persons to computer systems (Muller et al. 1997). To a major extent these activities are enforced by legislative constraints that exist in the US (e.g. the Americans with Disabilities Act (United States of America, 1990)) as well as in the European Union (European Commission, 2000), and in its member countries, like in Germany (Bundesrepublik Deutschland, 2006, Bundesrepublik Deutschland, 2002). However, a large amount of these realizations allow the user only to download particular forms, to print them, and

[1] This work is supported by the German Ministry of Research and Technology under grant FKZ 1771X07

Please use the following format when citing this chapter:

Kuhn, N, et al., 2008, in IFIP International Federation for Information Processing, Volume 272; *Human-Computer Interaction Symposium*; Peter Forbrig, Fabio Paternò, Annelise Mark Pejtersen; (Boston: Springer), pp. 119–128.

send it back to the governmental institution after some information has been inserted. Up to now, in many countries the use of printed documents remains the dominant means for the communication between governmental institutions and citizens.

While for the web based information systems accessibility aspects are often considered in E-Government platforms, for the procedure of forms completion in most cases it is necessary to process printed documents, yielding a point of media disruption which is difficult to handle for many users with particular handicaps.

In this paper we present an approach to build interfaces to governmental forms. Our main target is to develop tools to be able to cope with printed documents. However, some of the ideas can also be applied to electronic documents. Therefore, we exploit different computer science technologies e.g., from the fields of document analysis, language processing, and distributed systems to develop our solution.

The work presented here is embedded in the FABEGG project (Framework for modeling and acceleration of E-Government applications processes) and the GUIDO project (Generating user-specific interactive documents) which are both granted by the German Ministry of Research and Technology. In these projects we are concerned with the human computer interfaces to E-Government information systems. We deliver general guidelines how to build such interfaces, and develop generic concepts and modules that should be available for different platforms. We base our document representation on XML structures and implement communication by using web services. This guarantees independence from software and hardware platforms.

Our prototypical system used to demonstrate and to evaluate our work is called EG-VIP. In most cases when we speak about documents within this paper we have in mind governmental forms, which are used to provide and to maintain information that is necessary to run a governmental process. Such forms may be partially filled already, either by the authority itself or by other persons who interacted with the document before. In particular, when information from a citizen is required the authority usually sends a printed form which has to be completed and returned to the governmental unit where the next step in a process is executed. Thus, a form is often an instance of some form template to which individual data identifying the recipient and the corresponding process is added. This is e.g. the case for mail questionnaires in legal processes, real state affaires or tax proceedings.

2. System Architecture

Our main target is to develop interfaces to E-Government systems which are able to handle governmental forms. Moreover, we consider for this task users suffering from reading disabilities. On the one hand, these are people with visual impairments, ranging from colour blindness, over low vision to blindness. Another group of users we have in mind are those who suffer from dyslexia. All these users imply quite different requirements to the design of the interface and the interaction with an information system. In order to be able to economically develop adequate interfaces the system should adapt almost automatically to the actual user's needs. Figure 1 gives an overview of the system architecture we have chosen for that.

The EG-VIP system consists of different components. The core component is the GUIDO server in the lower left corner of the figure. It is basically a document database where all the documents of a governmental authority may be stored. A database entry for a document consists of different data. This comprises an image of the document which can be used to print it or display it on a screen. Furthermore, we expect to have a semantic description of the document, including information about the fields of a form, their type, or possible relations between them, which ones will be filled already with information and which ones have to be completed by the addressee. Also a link to the workflow management procedure in the authority should be present, i.e. a pointer to the description of the workflow where the document is used for. We call the universe of information associated with a governmental document or a form the Generic Document Structure. The document repository with all the generic document structures is built by the governmental agencies or associated business companies, e.g. the utility company or the water works.

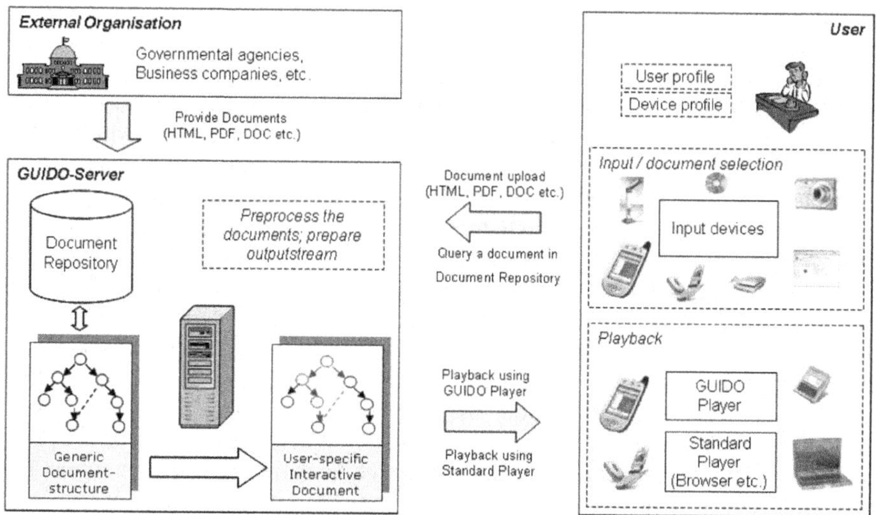

Figure 1. Architecture of the EG-VIP system

When a user interacts with the server site, the EG-VIP system derives an appropriate representation of the governmental documents for the user. We call this representation the User-specific Interactive Document. To compute the latter one from the generic document structure the system needs information about the user's sensory and physical capabilities and about his preferences. This will determine whether the server will run OCR or not, includes audio information, or computes transformations of the image which is actually processed. The choice will be based on the user profile, the actual bandwidth available for the client-server communication and on the features of the user device, which are given by the device profile. If the client computer has no speech synthesis program available the audio information has to be computed by the server. Bandwidth parameters can then determine the compression rates of the audio and the graphical information.

3. Constructing the User Interface

The right-hand side of figure 1 shows different devices with which a user may interact with an authority portal. This includes a desktop computer, or a smart-phone. Other devices may also be used, but the figure includes only devices which are able to start interaction with the server as well as to receive the corresponding answer. However, this assumption is not necessary and could easily be discarded.

At the beginning of the communication the user sends an image of the form he wants to work on to the server. He may take a picture either with the device itself (e.g. the camera within the mobile phone) or with a separate device, for example a digital camera or a scanner. Together with the image, the device and the user profile are sent. The device profile comprises information about the features of the device sending the request, e.g. the size of the display (height and width), or information about the software which is installed, for example whether an OCR program or a speech synthesis tool are available on the device. In addition, the service considers which software is used to present the result to the user. For example, this may be an EG-VIP specific software component allowing more interactions for the user compared to the situation when a general tool like a browser is used.

The user profile defines personal preferences for the visual or the audio result of the request. This may concern for example the color combination of the document which will be displayed by the device (e.g. black-white, black-cyan, cyan-yellow, or none). For the audio result the compression method or the sampling rate can be specified. Both, user profile and the device profile determine the result which is expected from the server. They may vary for the same user according to the situation or the location in which he currently is. For example, he may have installed an OCR-software on his desktop computer but not on his mobile device, or he prefers to receive only the audio output when he is traveling around, while at home he wants to receive a black-green document which he can enlarge and read himself. Both kinds of profiles are described in an XML-file which is transferred to the server as part of the service request.

When the server receives such a message it first has to analyze the document, i.e. the server tries to find an appropriate generic document structure within the document repository. For this task we currently use the result of an OCR process giving us specific keywords which are used to retrieve a document description. Other approaches, for example barcode recognition could also be useful for this purpose.

After the document has been successfully identified its generic document description is read which contains information how to process the image further, e.g. it describes which information can be expected to be already available in certain fields and hence, has to be extracted by the OCR process and which information should be requested from the user. The server packages all the relevant information into the answer it returns to the user client. In order to achieve platform independence and to be able to incorporate into our scenario all the devices shown in figure 1 we encode the answer into an XML file format. Figure 2 below shows an excerpt of such a file.

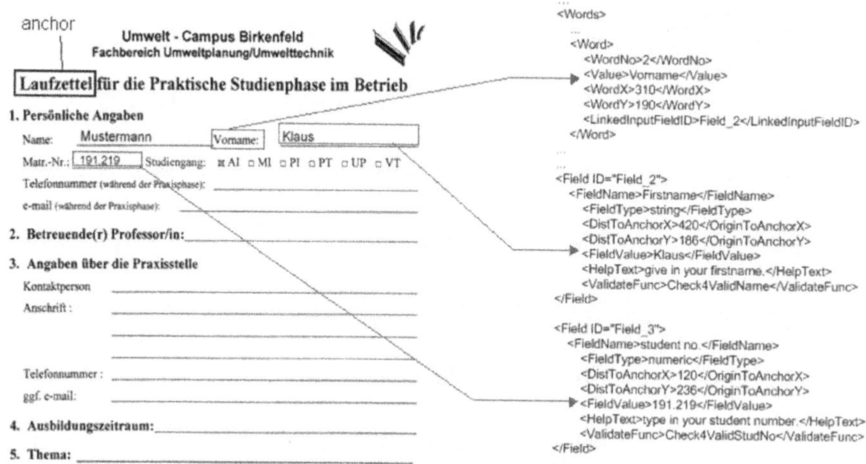

Figure 2. A Governmental form and its representation in the user-specific document

On the left-hand side of figure 2 you may find a form from our local university administration. It contains already the name, surname, and the registration number of a student. This information is extracted automatically from the document using the information which fields have to be processed in this and where they are located. This data is encoded in the XML-document as shown on the right-hand side of figure 2.

The structure of the result is almost the same in any use-cases. However, there are different ways to use this structure. In the following we want to discuss two scenarios.

3.1 Example 1: Desktop Client

Figure 3 shows a high level configuration of a user client system running on a desktop computer. It consists of a camera unit -shown on the right-hand side - that allows to capture printed documents and the display with speakers built-in. The camera unit enables easy document handling and fast document capture and moreover, is able to treat documents of larger size. Alternatively a scanner could be used. Such a user terminal could be placed for example in a major administrative department. In this case the speakers would be headphones due to privacy considerations.

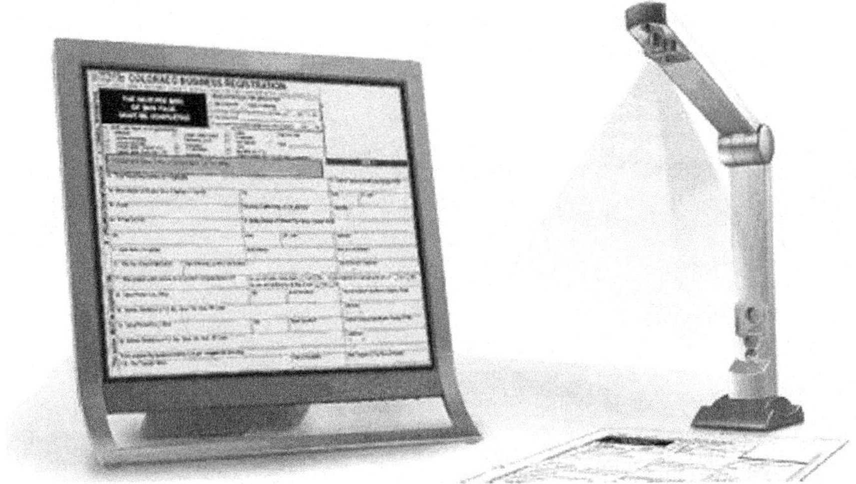

Figure 3. Desktop client with digital camera

The system can further vary according to the software which is installed. For example, we may have OCR-software, as well as speech-synthesis and speech recognition software locally available resulting in a highly autonomous system. This setting corresponds widely to the features of our first prototype which is referred to for example in (Kuhn et al. 2007a) and (Kuhn et al. 2007b).

In this environment it is possible that after the document has been processed by the local OCR software it can be read by using the speech synthesis software. The software component we use for speech synthesis also allows text reading in different velocities. While reading the text the system highlights the word which is actually spoken by drawing a (maybe coloured) rectangle around it. This enforces auditive understanding. With a pointing unit (e.g. a mouse or the finger when a touch screen is available) the user can start and stop reading at any arbitrary position. As the system is also able to process several common electronic formats like PDF or HTML, it is also an appropriate means for providing access to general electronic information systems.

With this configuration communication between the client and the server is only necessary when semantic information about the document is asked for. So the server can provide the client with type information about the fields on a form or with information related to a process where the form belongs to. In turn, this information can then be used to verify data from the OCR software which have not yet been recognized with sufficient confidence or to check user input for validity. Another important type of information which the server can deliver are help texts to guide the user through the process of completing the form and explaining to him which information is required in a concrete situation. The help texts are also present in the XML document of figure 2.

If we have a client at our disposal where less software features are installed the server is urged to provide the lacking information. Usually we encounter this situation when we use a client with less computational power, like a mobile phone or a smart-phone.

3.2 Example 2: Mobile Client

In 2006, Kurzweil Technologies Inc., an American manufacturer of assistive products and the National Foundation of the Blind presented a mobile phone equipped with OCR software which allows reading texts with that phone which are captured by the integrated camera (Kurzweil Technologies Inc., National Foundation of the Blind 2008). Recently, in cooperation with Nokia similar software has been developed and can be bought for a Nokia mobile phone, which is much lighter and more powerful than the original knfb reader.

This product yields a substantial improvement for the autonomy of humans with low vision when they are en-route away from the equipment in their home office. However, the solution is currently bound to single hard- and software system. The approach we present here overcomes this restriction and can in principle be run on any device which has available an internet browser. Nowadays this is true for the vast majority of the mobile phones which are being sold.

Figure 4 shows the representation of a German governmental form on a smart-phone emulator from the MS Visual Studio .NET development framework. We used this framework to implement the mobile version of our EG-VIP system. Many of the interactions between user and interface known from the desktop client are also available on this platform now.

Figure 4: Form representation on a mobile client.

As you can see by the box around the text, it is possible to step through a document word by word and get the word which actually has the focus visually marked while it is read. It is also possible to choose among different color combinations or to enlarge the image. Furthermore, the service can deliver the audio information which represents the spoken text of the document. For this the user may choose among different modes, i.e separate the audio information into different streams for each word, or for each paragraph or receiving a single stream.

Unlike the Nokia/Kurzweil approach we execute most of the computation on the server. To do so, we have encapsulated our formerly local application into a webservice which can interact with the mobile client over a network (e.g. the internet) answering HTTP based requests. One advantage of this solution is that on the server we can apply more powerful software tools, e.g. for the OCR and the speech synthesis. Thus, we expect much better results concerning the text recognition rate and the quality of speech.

The computational task of the client is restricted to the interpretation of the user-specific interactive document (cf. figure 1). As mentioned before this is an XML document where the client can find any data which is necessary to display a document and which the client can not (or does not want to) produce itself. In an extreme case the XML document can degenerate to an HTML document, containing for example only the document in a certain color combination or represented as an audio file which is played by a browser.

4. Conclusion and Outlook

The EG-VIP system provides humans with reading disabilities with much better access to information systems. In particular, access to printed documents, which are still important in business processes or in governmental applications, is substantially improved. Moreover, our solution can also be used in every day situations, e.g. when a visual impaired person stands at a bus stop and needs to read the schedule there.

We have further discussed our ideas to automatically adapt the output of a centralized information portal of a governmental authority to the special need of a citizen looking for information. The core of our system is the document repository where the different documents are maintained and the user specific interactive document is generated. The implementation of both a desktop client and client for smart-phones gives us strong support for the viability of our approach.

In our projects different partners are involved, among them a governmental authority and a German manufacturer of assistive technologies. Together with them we will integrate our approach for dynamically constructing the document interface and evaluate it in field tests together with citizens and handicapped users.

Currently, our main focus lies on the group of users with visual disabilities. This group of users will increase in number at least in many European countries due to the demographic development. A further possible target group are people with difficulties in reading comprehension, e.g. immigrants or dyslexic people. Studies from the OECD have stated that all over the world there are 800 million people who are not able to read or write. Even in the Western European countries the percentage of the inhabitants showing functional reading disabilities is estimated to be more than 15 percent (International Adult Literacy Service, 1998). Providing them with the automatic reading of documents and including in addition a speech recognition facility into the system will allow them to access to governmental processes more easily.

References

Bundesrepublik Deutschland, 2006. Allgemeines Gleichbehandlungsgesetz der Bundesrepublik Deutschland vom 29.06.2006, http://www.gesetze-im-internet.de/agg/

Bundesrepublik Deutschland, 2002. Verordnung zur Schaffung barrierefreier Informationstechnik nach dem Behindertengleichstellungsgesetz (Barrierefreie Informationstechnik-Verordnung - BITV) vom 17. July 2002

Edwards, A. D., 1988. The design of auditory interfaces for visually disabled users. In Proceedings of the SIGCHI Conference on Human Factors in Computing Systems (Washington, D.C., United States, May 15 - 19, 1988). J. J. O'Hare, Ed. CHI '88. ACM Press, New York, NY, 83-88.

European Commission, 2000. Gleichbehandlung Behinderter in Beruf und Bildung: Richtlinie des Rates 2000/78/EG vom 27. November 2000, ABl. L 303 vom 2. Dezember 2000.

International Adult Literacy Service, 1998, Report for the Organization for Economic Co-operation and Development (OECD), http://www.statcan.ca/english/Dli/Data/Ftp/ials.htm

Kuhn, Norbert; Richter, Stefan; Naumann, Stefan: Improving Access to EGovernment Processes. In: Khosrow-Pour, Mehdi (ed.): ManagingWorldwide Operations and Communications with Information Technology. Proceedings of the 2007 Information Resources Management Association International Conference (IRMA 2007) Vancouver (British Columbia), Canada (2007a). IGI Global, Hershey / New York 2007, pp. 1205-1206

Kuhn, Norbert; Richter, Stefan; Naumann, Stefan: Improving Accessibility to Business Processes for Disabled People by Document Tagging (Poster Presentation). Proceedings of the Ninth International Conference on Enterprise Information Systems (ICEIS 2007). Funchal (Madeira), Portugal (2007b), pp. 286-289

Kurzweil Technologies Inc., National Foundation of the Blind: Reader software for a mobile phone, http://www.knfbreader.com/index.php (15.02.2008)

Muller, M. J., Wharton, C., McIver, W. J., and Laux, L. 1997. Toward an HCI research and practice agenda based on human needs and social responsibility. In Proceedings of the SIGCHI Conference on Human Factors in Computing Systems (Atlanta, Georgia, United States, March 22 - 27, 1997). S. Pemberton, Ed. CHI '97. ACM Press, New York, NY, 155-161.

Richter, S., 2003: Design and Implementation of a communication module for blind and visually impaired humans. Diploma thesis, Birkenfeld.

Spiegel Online: Handy für Blinde liest gedruckte Texte vor. http://www.spiegel.de/netzwelt/mobil/0,1518,532014,00.html (15.2.2008)

United States of America, 1990. Public Law 101-336, 1990. Text of the Americans with Disabilities Act, Public Law 336 of the 101st Congress, enacted July 26, 1990.

W3C, 2006: Guidelines and resources from the World Wide Web Consortium (W3C). www.w3.org/WAI/

Communicability in multicultural contexts: A study with the International Children's Digital Library

Clarisse Sieckenius de Souza, Robin Fred Laffon, Carla Faria Leitão

SERG – Semiotic Engineering Research Group, Depto. de Informática, PUC-Rio, Brazil, {clarisse, rlaffon, cfaria@inf.puc-rio.br}

Abstract: This paper presents some contributions of semiotic engineering to the identification of cultural issues involved in the design and evaluation of multicultural systems (i.e. systems designed for users from different cultures). We carried out a communicability evaluation of the International Children's Digital Library. Participants of test sessions had different nationalities and spoke different native languages. We paid special attention to communicability problems stemming from *language understanding* and *language use* issues. Our goal was not to make generalizations from our findings, but rather to uncover and understand some of the users' experience with multicultural systems. In addition to this understanding, we have gained relevant insights to inform the cultural adaptation of ICDL to the Brazilian context, and believe that they can be useful in other cultural contexts as well.

Keywords: Semiotic engineering, communicability evaluation, cultural issues.

1. Introduction

The dissemination of Information and Communication Technology (ICT) has brought about considerable cultural changes. ICT users come from ever more diverse cultural backgrounds, and technology continually fosters new cultural exchange (Castells, 2001). Stimulated by this scenario, the HCI community has set out to investigate design and evaluation methods and techniques that can improve the users' experience when cultural issues are involved. Cultural usability studies propose design strategies centered on both internationalization (to develop sufficiently generic interfaces, suitable for users from different cultures) and localization (to develop customized interfaces for specific cultures) (Aykin, 2005; Dray, 1996; Marcus, 2005).

Please use the following format when citing this chapter:

de Souza, C.S., Laffon, R.F. and Leitão, C.F., 2008, in IFIP International Federation for Information Processing, Volume 272; *Human-Computer Interaction Symposium*; Peter Forbrig, Fabio Paternò, Annelise Mark Pejtersen; (Boston: Springer), pp. 129–142.

These studies have been particularly important for multicultural systems whose users do have different cultural origins, but whose primary purpose does not include cultural exchange. Such is the case of web search systems, for example, where internationalization and localization have brought concrete improvements for the users' experience. However, when it comes to systems whose main purpose is precisely to support social exchange among people from different cultures, internationalization and localization fall short of offering much needed solutions. These systems aim to *communicate culture for people from different cultures*. Such is the case of websites designed to support travelers' exchange of stories and experience from all over the world. In a context like this, localization is barely useful because it prevents users from experiencing *cultural diversity*, which is ultimately what they are looking for. And internationalization may arguably be subtracting (for sake of *generic solutions*) some specifics ('the *local flavor*') that have a great value for users. A useful metaphor, for sake of introduction, is to say that cultural exchange systems should provide the equivalent of a 'tourist information service'. They should allow users to experience diversity and local values and practices, but at the same time they should prove cultural *scaffolds* (brief informative descriptions and narratives, tips.) for visitors and newcomers.

This paper presents some preliminary contributions of semiotic engineering (de Souza, 2005), a semiotically-inspired theory of human-computer interaction (HCI), to conceptual design of multicultural systems. These contributions stem from an evaluation study carried out with ICDL – the International Children's Digital Library. ICDL is a specialized public digital library on the Internet, with children's books from throughout the world. An initiative of researchers from the University of Maryland, its mission is "to support the world's children in becoming effective members of the global community - who exhibit tolerance and respect for diverse cultures, languages and ideas" (ICDL, s/d). Among many design challenges, ICDL must support multicultural human-computer interaction.

In semiotic engineering, the main function of computer systems interfaces is to bring designers and users together at interaction time, and to communicate the designers' vision to users through computer-human interaction. This integrative view of a continuous communication process expresses – directly and/or indirectly – the goals and purpose of technology, the choice of interactive strategies and tactics, and in the case of cultural exchange systems, a characterization of different cultures, how they are approached by the designers, and also how users can navigate from one culture to the other.

ICDL-Brasil (ICDL-Brasil, s/d) is a binational cooperation project, whose aim is to make ICDL useful and usable for Brazilian children, in an attempt to encourage them to read and write more and better. A vast share of the Brazilian population is plagued by functional illiteracy. According to recent demographics (INAF, 2005), only 26% of Brazilians between ages 15 and 64 are fully capable of reading and writing. The first step in the project is to assess how users immersed in a Brazilian context interact with the current ICDL interface. Among the tools we have used to this end is a qualitative method proposed by semiotic engineering: the communicability evaluation method, CEM (de Souza, 2005; Prates et al., 2000).

Unlike in most previous ICDL evaluations (e.g. Bilal, 2007; Druin, 2005; Hutchinson et al., 2005), our participants were not children, but adults. Although this may sound surprising, the role of adults in getting young children interested in reading has been shown to be crucially important (INAF, 2005). Hence, a critical requirement of ICDL-Brasil is that adults using ICDL with children feel comfortable and stimulated.

Our study has produced an in-depth perspective into some of the interpretive processes occurring in the context of typical interactions with ICDL. Our major findings are:

1. ICDL should provide increased support for multicultural navigation in order to increase the sense of safety and comfort of users as they move across different cultural settings.
2. As a multilingual site, ICDL should pay closer attention to the various linguistic materials in the system (*e.g.* the interface language and the various languages in which the books are written), in order to prevent serious interactive breakdowns.
3. ICDL design would probably benefit from separating linguistic and pragmatic issues in multicultural interaction, especially with respect to different forms of manipulating and reading books in Eastern and Western cultures.

In the following we briefly present the gist of semiotic engineering. Next, we introduce CEM. Then, we present our case study. Finally, we discuss our findings in view of related research and present our conclusions.

2. Semiotic engineering: communicability and culture

Semiotic engineering (de Souza, 2005) is a semiotic theory of HCI that views human-computer interaction as a particular case of metacommunication (communication of/about communication). Assuming that software is an intellectual artifact, the result of rational decisions and choices, and is designed to achieve certain purposes and effects in order to benefit and/or please its intended users, semiotic engineering stresses the fact that software requires appropriate interactive presentation and introduction. In accordance with this theory, the natural way of promoting good encounters and experiences with technology is to tell users, at interaction time, what they need to know in order to take the best out of it. But the way to do so can vary widely.

A key concept for semiotic engineering is communicability. Communicability is the distinctive quality of interactive computer-based systems that communicate efficiently and effectively to users their underlying design intent and interactive principles (Prates et al., 2000). As metacommunication artifacts, systems communicate a message that is elaborated by designers (senders) and interpreted by users (receivers). Together at interaction time as interlocutors of a communication process, designers and users negotiate meanings. And one of the keys to negotiating meanings is the designers' ability to communicate their intent (de Souza, 2005). The essence of this communication is the answer that a designer (or spokesperson for a design team) can give to three sets of questions:

1. Who are the users of the product I have designed? What do they want or need to do? In which preferred ways? And why?

2. What is this product I have designed? What can it do for its users? How?

3. What kinds of interactions, conversations, can or should users have with it? What range of effects can these conversations achieve? Are they consistent with who my users are, and with the needs and expectations that they have?

Put together, the answers to the above questions compose a top-level metacommunication message that is sent from designers to users, through the system's interface. Users unfold and receive this message, as they interact with the system. Thus, a system's interface is at once a message and an interlocutor, in a twofold communication process. Users interact with the message in order to get it, fully. Since this interaction is itself a process of exchanging messages, the interface must speak for the designers at interaction time – be "the designer's deputy", as proposed by the theory.

In order to illustrate metacommunication, we can use an instance from Google Brasil (http://www.google.com.br), focusing on the internationalization and localization strategies they use. In Figure 1 we see Google's interface for Brazilian users, which is very similar to that of North American users (same content, same interaction patterns, same sign structuring). To this internationalization strategy, designers have added localization ones (a separate URL, Portuguese translation). A single English language sign is preserved – a link to Google in North America. Search results are also localized: Brazilian domains are considered more relevant than non-Brazilian ones, even when users are searching for a term in English.

Figure 1 Google Brasil homepage

Now, what are the designers telling us (receivers of their message at interaction time)? The Google team communicates that users from certain cultures, like Brazil in this case, can benefit from an interface in Portuguese, with its own specific URL ('.br' suffix at the end of Google's generic 'www.google.com' locator). They also communicate their belief that Brazilian users prefer to access information in Brazilian Internet domains. However, if they wish to use the international Google version, they can reach it quite easily, by clicking on "Google.com in English".

This small glimpse into the reconstruction of the metacommunication message from Google Brasil helps us see what the designer-to-user communication process is and how it is achieved through the designer's deputy mediation at interaction time. It also helps us illustrate a potential communicative breakdown with the Google Brasil interface. When searching, Brazilian users may not realize that results from a search carried out in the Brazilian site do not come in the same order as results from the same search in the North American site. In other words, this localization effect is not clearly communicated. Only by comparison with searches in other Google sites does the user get the message.

Moreover, because the Google Brasil example so clearly directs us to cultural issues involved in the design of a very simple interface, it helps us understand why and how semiotic engineering is so tightly connected with culture. European semiotic theories (Eco, 1976) view communication as a cultural process in which we express ourselves through signs. Interpreting a message involves giving meaning to signs, signs that are most often established by virtue of cultural conventions. Thus, when we say that in semiotic engineering HCI is a particular case of metacommunication, we are actually implying that everything that constitutes an HCI issue is, in essence, a cultural issue.

Although the concept of culture is not explicitly defined in semiotic engineering, it is inexorably bound to this theory's basic ontology. Umberto Eco (1976), for instance, defines semiotics as "the logic of culture", and Clifford Geertz (1973) argues that the nature of culture is essentially semiotic. Culture, in this perspective, is a shared signification system that makes our behavior and the world around us meaningful; it shapes the way we interpret reality. So, metacommunication is a cultural process by which designers express a considerable extent of their world view.

3. The Communicability Evaluation Method

The Communicability Evaluation Method (CEM) (Prates et al., 2000) is a qualitative evaluation method developed to capture communicability problems, which are ultimately related to metacommunication issues. CEM privileges the reception of the designers' metacommunication, and produces not only an identification and explanation of problems, but also information for redesign.

CEM introduces empirically-grounded pieces of knowledge in the collection of interpretive resources that designers can use to make sense of observed reality. Like other qualitative methods (Denzin and Lincoln, 2003), CEM emphasizes the diversity and richness of meanings involved both in users' and designers' interpretations of empirical reality. Each user's test session is unique and also bears many possible interpretations. Actually, the interpretation of several participants' results does not promote a consensual and convergent set of replicable parameters, but rather a set of empirical knowledge that stimulates the designers' critical and creative reflection about the metacommunication.

CEM is carried out in four steps: test preparation, tagging, interpretation and semiotic profiling. It involves observing users during interaction, analyzing these observations, exploring the application further, and finally explaining identified metacommunication problems, along with suggestions for redesigning the application.

The test preparation step is similar to typical user testing preparations, except that it requires that the evaluators elaborate an explicit description of the metacommunication message, and relevant test scenarios, with appropriate contextual cues for the analysis of metacommunication.

Tagging is the heart of CEM. Once user sessions have been recorded, evaluators watch the recorded movies and "put words on the users' mouths", in a kind of after-the-fact reconstruction of a verbal protocol. This is not a free text annotation of the recorded interaction, but a principled association of thirteen technically defined tags to those portions of the movies where evaluators detect communicability problems. For sake of illustration, among the communicability tags used in CEM, we have: "Where is it?"; "Where am I?"; "Oops!"; and "I can't do it this way.".

"Where is it?" is tagged in contexts where user knows what she is trying to do but cannot find an interface element that will tell the system to do it. She typically navigates through web pages or opens and closes dialogs, looking for that particular element.

"Where am I?" is tagged when the user is telling things to the system that would be appropriate in another context of communication, but not in the current one. Symptoms may include trying to select objects that are not active in the current context, trying to interact with signs that are output only, and so on.

"Oops!" is tagged when the user makes an instant mistake, and immediately tries to correct herself. A typical symptom of "Oops!" is to undo the faulty operation triggered by miscommunication.

Finally, ***"I can't do it this way."*** is tagged when, while trying to achieve a goal or sub-goal, the user engages in a several-step sequence of operations, but suddenly realizes that this is not the right thing to do. So, she abandons that long sequence, and takes a different path.

The interpretation step amounts to determining how successful the designers' communication is. Success is associated to the absence (or insignificant amount) of communicative breakdowns. Evidence from the tagging step helps evaluators decide on the quality of such communication. The following factors help the evaluators identify and understand communicability problems:

1. How often, and in which particular context, each type of tag appears;
2. The occurrence of tagging patterns;
3. Regular associations of tag types or sequences with problems in establishing communicative goals; and
4. When using additional evaluation methods, a correspondence between the locus of tag occurrence and that of problems indicated by the other methods. At the end of this step, evaluators should be able to tell when, where, how and why observed users were unable to: express what they meant; understand the system's expressions; choose the right way to communicate their intent; assigned the right meaning to what the system was communicating; or formulate a communicative intent altogether.

Semiotic profiling is the final step in CEM. The goal at this stage is to identify and explain problematic interaction design, and to inform redesign. They reconstruct the designer-to-user metacommunication message based on evidence provided by tests with users and on further exploration and inspection of the application

CEM thus achieves two important results. First, it gathers relevant evidence for redesign. Second, CEM expands the evaluators' and the designers' knowledge about HCI. For design, in particular, semiotic engineering explanations for communicative breakdowns can be the seed to more elaborate reasoning and decision-making when choosing between design alternatives, or when generating alternatives themselves. This is why, CEM is an epistemic tool. Its purpose is not to dictate solutions to a problem, but to support problem-solvers in naming and framing design problems, in generating solutions, and evaluating them ones against others (de Souza, 2005).

In the next section we show examples of CEM results when applied to ICDL.

4. The ICDL Study

One of the main practical goals of ICDL-Brasil (ICDL-Brasil, s/d) is to elaborate a culturally-adequate interface for ICDL, so that it can be productively used by Brazilian children and adult tutors. Thus we started the project with a communicability evaluation, looking for breakdowns related to cultural issues.

In the preparation phase we selected the cultural dimensions that would guide our study: language and pragmatics. Language, in this context, refers to the diversity of tongues, and pragmatics refers to the diversity of behavioral practices and attitudes involved in language use and in dealing with linguistic objects (Leech, 1983). There are several reasons for our choice. Firstly, the fact that languages encode (and thus reveal) culture; that is, they are primary signification systems (Geertz, 1973). Secondly, ICDL is a collection of instances of language use – many languages are spoken in and with ICDL (through books and interfaces). And finally, an important part of the ICDL mission is to foster "tolerance and respect for diverse cultures, languages, and ideas" (ICDL, s/d).

Next step, still in the preparation phase, we defined the participants' profile. Because of the leading role of adults in promoting children's literacy (INAF, 2005), we decided to start with a group of young adults. Participants were six students with experience in intercultural exchange programs, who could speak at least one foreign language. All of them had had previous experience with children (e.g. as teachers, baby sitters, and so on). The group was a bi-national one: three French and three Brazilian students. Note that there were no participants from the same culture as ICDL developers (North American). The last step in the preparation phase was a careful inspection of ICDL, from which we reconstructed the designers' metacommunication message and derived a test scenario.

At test sessions, all participants received the following assignment: *"Suppose that you are a teacher or educator working with a group of 8-year olds. Because you loved your intercultural exchange experiences in the past, you want to light up their interest and curiosity about foreign cultures. Your goal is to search this website and choose a book to show them. You want to be original: you want to pick up a book in a language that they really don't understand, in this case in Persian or Farsi. This should trigger their imagination. It could be about music or poems."*

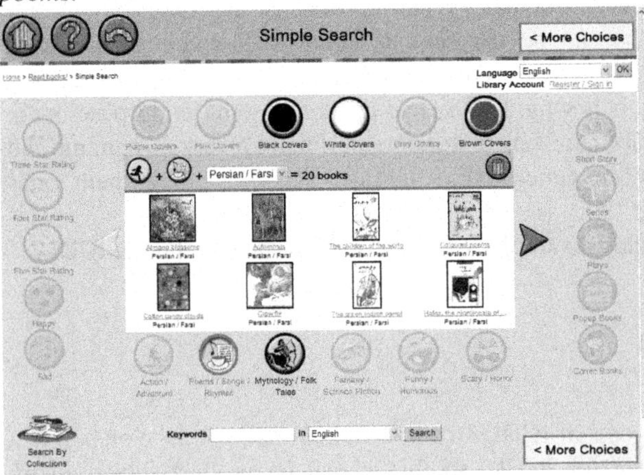

Figure 2 A simple search for books about poems / songs / rhymes in Persian/Farsi

The task was fairly open-ended, and explicitly addressed the multicultural character of ICDL. In Figure 2 we see a snapshot of a simple search returning the kinds of books that participants might choose. Using the simple search for the task is very convenient. The set of books to choose from can be obtained in only four steps: clicking on the appropriate age group; selecting the Persian/Farsi language; asking for "more choices"; and clicking on type "poems / songs / rhymes".

However, it took participants more than 8 minutes on average to finish the test. None of them searched for books in the way suggested in Figure 2, but all except one (who chose a book in Serbian) found an appropriate book.

Tagging was the next step and, below, we present the recurrent tags of our study.

Oops!: All participants except one were deeply confused by the language choice when searching a book (see Figure 3). They took "language" to mean book language, not interface language. Thus, when they set it to Persian/Farsi, they got into serious trouble. None of them could understand Persian/Farsi, so this was a mistake they had to correct. However, this "Oops!" was an expensive one, as will be seen below.

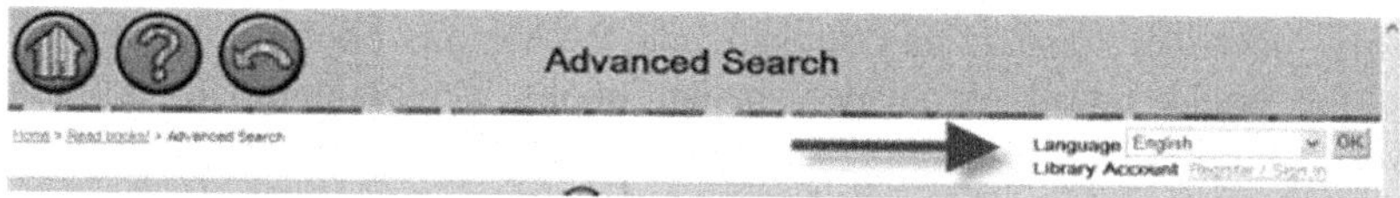

Figure 3 Choosing the interface language in ICDL

Where is it?: When ICDL users first access the website, all communication is in English, which is perfectly acceptable. Our tests provided evidence that when users inadvertently set the interface language to Persian/Farsi, they immediately tried to reset the interface language. They started looking for an interface element (a sign) to communicate they wanted to undo the faulty language setting, to set it back to the English default. A common pattern in most recorded sessions was to navigate back to the homepage (in English), hoping that this action would lead back to the default language interface. In other words, while receiving metacommunication, they took "homepage" to mean the same as "restart", which is a misinterpretation of the concept of homepage in web navigation that concomitantly achieves some procedure. Another pattern was trying to set the language back to the desired value. The problem is that participants had to know how to say "English" in Persian/Farsi. Since they didn't, interaction degraded into a lengthy quest for comprehensible communication.

Figure 4 Setting the interface language back to English in Persian/Farsi

Where am I?: The inversion of navigation hints in the Persian/Farsi interface confused most participants. For example, advancing (going forward) and returning (going backward) were signified the other way around for Westerners. So, whenever they used arrows to mean "back/previous", they were actually telling the system to move "forward/next". The problem also appeared when the interface language was right for the user, but the desired action was to preview a book. When a Persian/Farsi book is previewed, the pages are displayed from right to left. This caused great confusion because it took some participants a while to realize why what they thought should be the book's first book page was an arbitrary page spatially located on the top-left corner of the overviewing area (see Figure 5).

I can't do it this way.: The location search is primarily a search by continent, and not by country as the link "books by country" suggests. Only when a particular continent is selected can the search be narrowed to a country (e.g. Asia and Middle East > Iran). This strategy cannot be combined with other criteria, such as content and age for instance. Thus, participants that chose this strategy could only meet their goal if they looked into the various books' metadata or browsed the books trying to decide whether books were appropriate or not. Moreover, a book "from a country" in ICDL is not necessarily written in this country's official language. Thus, participants who tried to use this kind of search either switched to another strategy after a while, or spent a considerable amount of time browsing books till they found one they liked.

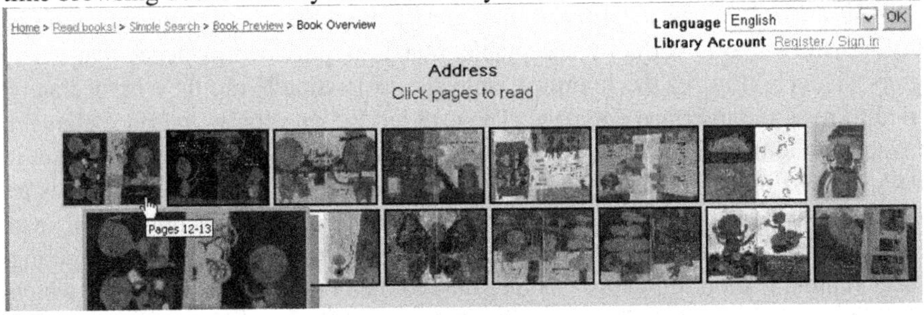

Figure 5 English interface overview page of a book in Persian/Farsi

After tagging, we interpreted results. Table II summarizes our main findings. The first column indicates a communicability problem. The second indicates whether we take it ("✓") as a cultural problem related to language. And the third column indicates whether we take it as a cultural problem related to pragmatics.

Some problems are pragmatic, rather than linguistic, in a strict sense. For instance, a native speaker of French may not understand Finnish. Yet, if this person picks up a book in Finnish, the order of pages and direction of reading will be intuitive, even though the words can't be understood. Contrastively, if this person cannot speak Hebrew, and picks up a book in Hebrew, she may or may not realize that the direction of reading is reversed. Notice that, just as with Finnish, she doesn't understand the language. However, there is a sense of familiarity in manipulating books, identifying printed letters, and so on, that is much stronger with Finnish than Hebrew. This familiarity stems from practices and attitudes related to language use and dealing with linguistic objects regardless of how well one can understand one or another idiom.

Table 1 A summary of CEM results

Communicability Problems	Language	Pragmatics
"Books by country" is an ambiguous expression (books by country = books by continent; a country's book may not be written in the country's language).	✓	
Out-of-sync metadata translation may expose users to unexpected changes of language during interaction (interface in Portuguese, metadata in English).	✓	✓
Users could not find how to reset the interface default language.	✓	✓
Users were confused with the order of reading / navigation when exposed to Persian/Farsi interface and book previewing.	✓	✓
Users are confused by "interface / keyword / book" language settings.	✓	
Multilingual representations and multilingual instances of books can be easily confused (translated metadata ≠ a translated version of the book).		✓

Other problems are strongly related to language, and only weakly related to pragmatics, like translation problems. For example, there was some confusion about books, themselves, and books metadata. The metadata may appear in two languages (e. g. English, the default, and French). When users see a book cover on screen labeled "Mushrooms in the rain", and later the same book cover is shown with the label "Champignons sous la pluie", they cannot tell whether these are two books, or the same book, whose title is translated into a foreign language.

Our interpretation of tagging led us to relevant insights in the semiotic profiling stage. We concluded that the main communicability issues in ICDL are three.

First, as a multicultural online environment par excellence, ICDL should provide increased support for multicultural navigation. Users should not lose sight of their native cultural markers when they move across different cultural settings. The presence of such markers (e.g. their native language, their homeland or starting point in the multicultural journey, etc.) could increase their sense of safety and comfort, and encourage them to take even wilder journeys in the ICDL globe. Our scenario, where an Eastern language and culture was chosen as a target for Western users, was deliberately chosen. It is in tune with the ICDL design vision. However, in the absence of cultural markers, users experienced serious problems with the interface.

Second, as a multilingual site, ICDL should separate different dimensions and classes of languages in its interface, and communicate differences very clearly. In the analyzed interface, all these languages were bundled together: the interface language, the original book language, the languages into which a book is translated, and the metadata language.

Finally, decoupling linguistic from pragmatic issues in multicultural interaction may in fact benefit design. The problem with the direction of reading and book pagination is an important one. When reading a book, the user is supposed to understand the book language. Thus, within the scope of a visualized page, the direction of reading is of course affected by the particular graphical encoding adopted by that language. Hence, Persian/Farsi text can only be read right to left. However, book manipulation has to do with cultural practices that are language-separable. Our inspection of ICDL has shown that the direction of reading in ICDL can be very confusing. Not only can users find the situation depicted in Figure 5, but speakers of languages encoded from right to left may even be more confused when they browse pages of books encoded from left to right – the meaning of arrows in pagination is difficult to figure out. Decoupling language from book manipulation allows designers to think of users' intuitions about book pagination, and design better browsing strategies for books written in languages unknown to the user.

4. Discussion and Conclusions

The use of CEM as an evaluation tool in the study of ICDL has yielded useful results. The need for cultural references to orient ICDL users as they move across cultural boundaries is in line with Barber and Badre's findings about the role of cultural markers in improving the usability of international websites (Bilal, 2007).

Likewise, much of the problems we have treated as linguistic have been dealt with in cross-cultural research by Bourges-Waldegg and Scrivener (1998), although as generalized meaning and representation problems. The authors concede, however, that "particular linguistic representations [...] are a design issue only if they become an obstacle to understanding." (p. 301). And this is precisely what CEM has allowed us to verify in the ICDL interface.

We also share many of their views. For example, it is our common belief that cross-cultural communication in natural contexts is probably less problematic than culturally-oriented HCI research may lead us to believe. Even coming from different cultures, people are naturally aware that signs may be interpreted differently by foreign interlocutors. This multicultural awareness can not only be sustained, but also increased, by culturally-informed HCI design.

As mentioned before, CEM is a qualitative method, whose aims is not to generate universal knowledge, predictive or prescriptive. CEM results are always contingent to a specific context, in our case the specific ICDL interface with which our participants interacted. However, CEM's epistemic nature allows results to become part and parcel of a design knowledge base. Thus, semiotic engineering explanations for communicative breakdowns can be the seed for more elaborate reasoning and decision-making when choosing between design alternatives, or generating new alternatives. Hence, we believe that the findings in this study are relevant pieces of knowledge that can be used to make sense of others multicultural environments,as shown next.

The first finding refers to the presence of native cultural markers – the native language, the homeland or the starting point on multicultural journeys. The Google Brasil example in section 2 discussed the presence of such marker, represented as a single link in a foreign language ("Google.com in English"). The second finding refers to a separation between interface language and various domain languages. Here again we can go back to the Google Brasil example and see that although Brazilian users may be happy to see the interface in Portuguese, they may not be happy to realize that search results are influenced by the assumption that items in Brazilian sites (in Portuguese, too) are more relevant to the user than items in other sites across the world. The problem stems from equating language and culture, whereas language is actually an element of culture (though possibly the most important one).

Finally, our third finding stresses the importance of treating linguistic and pragmatic issues as different focal points in the design process. The manipulation of linguistic objects is especially relevant, for example, in the design of online manga reading sites. Manga are Japanese cartoon-like stories that, in printed form, are read according to Japanese reading practices, in the opposite direction of Western cartoons. Various online reading sites for manga (e.g. http://www.readmanga.com) explore the possibilities of computer technology precisely to help Western readers follow the stories more comfortably, without the disorientation that results from reading manga in print. Unlike what happens in ICDL, manga online readers typically strive to help Western users by adopting a Western pagination and book manipulation standard (see figure 6).

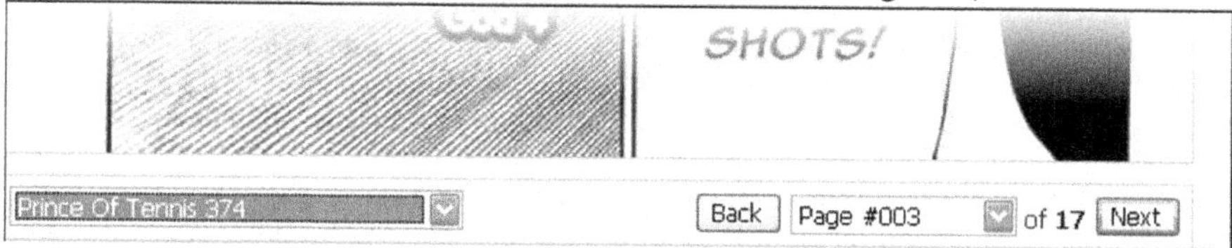

Figure 6 Interface of a manga viewer (http://www.mangavolume.com)

Based on these few examples, we consider that our results can contribute to support problem-solvers in naming and framing cultural design problems and in generating solutions for the design not only of ICDL but also of multicultural environments.

Acknowledgments This research is financially supported the National Council for Scientific and Technological Development (CNPq), Brazil, and is carried out in international cooperation with the ICDL research team at the University of Maryland, College Park. The authors would like to thank all the test participants.

References

Aykin, N.: Usability and internationalization of information technology. Lawrence Erlbaum, Mahwah (2005).

Barber, W, and Badre, A.: Culturability: The merger of culture and usability. Proceedings of the 4th Conference on Human Factors & the Web (1998). http://zing.ncsl.nist.gov/hfweb/att4/proceedings/barber/index.html. Accessed Jan 2007.

Bilal, D., and Bachir, I.: Children's interaction with cross-cultural and multilingual digital libraries: Understanding interface design representations. Information Processing and Management. 43: 47–64 (2007).

Bourges-Waldegg, P., and Scrivener SAR Meaning, the central issue in cross-cultural HCI design. Interacting with Computers. 9:287-309(1998).

Castells, M.: The Internet galaxy. Oxford University Press, Oxford (2001).

de Souza, C. S.: The Semiotic engineering of human-computer interaction. The MIT Press, Cambridge (2005).

Denzin, N. K., and Lincoln, Y. S.: The Sage Handbook of Qualitative Research. Sage, Thousand Oaks, California (2005).

Dray, S.: Designing for the rest of the world: A consultant's observations. Interactions.. 3(2), 15-18 (1996).

Druin, A.: What children can teach us: developing digital libraries for children with children. Library Quarterly 75(1): 20-4 (2005).

Geertz, C.: The Interpretation of Cultures. Basic Books New York.

Eco, U.: A theory of semiotics. Indiana University Press, Bloomington (1976).

Hutchinson, H., Rose, A., Bederson, B., Weeks, A., and Druin, A.: The International Children's Digital Library: A case study in designing for a multi-lingual, multi-cultural, multi-generational audience. Information Technology and Libraries, 24(1): 4-12 (2005).

ICDL: The International Children's Digital Library. (s/d). http://www.icdlboorks.org Accessed Jan 2007.

ICDL-Brasil: Uma biblioteca internacional digital para crianças brasileiras. (s/d). http://www.icdl-br.inf.puc-rio.br Accessed Jan 2007.

INAF 5o Indicador Nacional de Alfabetismo Funcional: Um diagnóstico para a inclusão social pela educação. Instituto Paulo Montenegro, Ação Educativa, São Paulo (2005).

Kaplan, N., and Chisik, Y. Reading Alone Together: Creating Sociable Digital Library Books, In Proceedings of IDC 2005. New York: NY. ACM Press. pp. 88-94 (2005).

Leech, G. N.: Principles of Pragmatics. Longman, London (1983).

Marcus, A.: User interface design and culture. In N. Aykin (Ed.) Usability and internationalization of information technology. Mahwah: NJ. Lawrence Erlbaum. pp. 51-78 (2005).

Prates, R. O., de Souza, C. S., and Barbosa, S. D. J.: A method for evaluating the communicability of user interfaces, Interactions 7(1): 31–8 (2000).

Searle, J. R.: Speech acts: An essay in the philosophy of language. Cambridge University Press, Cambridge (1969).

Facing the digital divide in a participatory way – an exploratory study

Elaine C. S. Hayashi[1], M. Cecília C. Baranauskas[2]

[1]Instituto de Computação – UNICAMP, Brazil, elaine.hayashi@gmail.com
[2]Instituto de Computação & NIED – UNICAMP, Brazil, cecilia@ic.unicamp.br

Abstract: One of Brazilian's grand challenges in computer science research concerns a "Participative and universal access to knowledge for the Brazilian citizen". In order to develop design solutions to address this challenge, we first need to understand these citizen's abilities and the context in which they are immersed. For that, we have been conducting practices actively involving a group of representatives of the diversity of users we have in the population. This paper presents the first results of this investigation, pointing out some lessons learned so far regarding the relationship they have with Information and Communication Technology (ICT) and how they make sense of different models of interaction to accomplish a simple task related to the exercise of citizenship. Among other findings, we were able to notice how their previous experience reflects on their behavior and the benefits of using an avatar in future systems.

Keywords: digital divide, accessibility, participatory design, universal design

1. Introduction

According to Drucker (1993), the main resource of our society in the future will be knowledge. Terabytes of information are digitally available and this amount grows everyday, but it is not everyone who is able to access it. The information is there and the means to spread it are also available through Information and Communication Technology (ICT). Some authors say the barriers to access knowledge do not have a technological nature, but rather social and economical ones (Varian, 2005), (Information, 2006); illiteracy for example is one of these barriers. In the reality of many developing countries, we have to consider both types of illiteracy, the literal and the digital illiteracy.

The Unesco's World Report (UNESCO, 2005) states that ICT creates conditions for the emergence of knowledge societies and these societies are a source of development for all. But for this society to arise, it is necessary to bridge the gap and diminish digital divide. In order to achieve that we need to develop systems that can be accessed by everyone. As stated by Shneiderman (2000), universal access initiatives should address at least three main issues: user diversity

Please use the following format when citing this chapter:

Hayashi, E.C.S. and Baranauskas, M.C.C., 2008, in IFIP International Federation for Information Processing, Volume 272; *Human-Computer Interaction Symposium*; Peter Forbrig, Fabio Paternò, Annelise Mark Pejtersen; (Boston: Springer), pp. 143–154.

(user with different skills, knowledge, age, gender, disabilities, literacy, etc), technology variety (support to a broad range of hardware, software and network access) and gaps in user knowledge (the difference between what users know and what they should know).

In a country that has continental dimension and 189 millions of inhabitants, the variety of its population could be huge. While 86% of Brazilian's adult population is considered literate (WorldBank, 2006) only 26% is actually capable of understanding a simple written text (Ribeiro, 2003), 14% has some kind of physical disability and 79% (of the population with 10 years old or more) have never accessed the internet (IBGE, 2005).

In spite of the supremacy of some computer hardware and software, it is necessary to think of solutions that can be accessed through any platform, i.e., that can be reachable by anyone. But much more than that is the challenge of designing interface and interaction solutions to reach the diversity of the population.

The Brazilian Computer Society's (SBC) challenge #4 dares us to think of and research about "Participative and universal access to knowledge for the Brazilian citizen". This is one of the five Grand Challenges in Computer Science Research for the years 2006 - 2016 that resulted from an event in 2006 sponsored by SBC (SBC, 2006).

Today there are public and private initiatives in the country to provide universal computer access, but most people do not make sense of possibilities brought by computers and internet. What kind of system could be developed that would work as a source of knowledge and would be appealing to the Brazilian's less favored population?

According to Melo and Baranauskas (2006), the accessibility recommendations are not enough to handle all the complexity that the *Design for all* demands in the scenario we have. To start facing the challenge, we decided to work directly with the user, who should have an active role in the design process. Participatory Design (PD) methods, tools and techniques have been successfully adopted (Rocha and Baranauskas, 2003), (Kensing and Blomberg, 2004),(Muller, 2002) in this sense: a design with, for and by the user (Lanzara, 1983). Including the user in the design process is vital to make sure we are creating systems that make sense and that are part of the users' context of life.

At the same time, we have noticed in recent years the amazing success that sites like Orkut are having in Brazil especially with the young people. Ramachandran et al. (2007) states that it is clear that "social networks play a dominant role in influencing the adoption and the use of ICTs". That gives us a hint that online social networks may be the means we were looking for to promote the digital culture in this population.

Within this context of universal and participatory access to information, we have been conducting participatory practices with a selected group of people in Campinas – a city in São Paulo State – as part of a project of a broader scope in which the objective is to develop online social networks that are inclusive and meaningful for Brazilian citizens. In this paper, we describe preliminary findings regarding activities we started within this community to investigate possible design solutions. The goal in the activities we are going to report in this paper was to recognize and better understand the relation this group has with ICT, how they interact and make sense of it. We also investigated how they respond to the different forms of media: text, sound and images. As a result, we present here preliminary findings and first lessons learned from this work.

The paper is organized as follows: the next section gives details about the scenario and group of participants in the study, describing the practices carried out. Section 3 presents preliminary results of analysis. Section 4 discusses some lessons learned from this experience and Section 5 summarizes a conclusion.

2. The Participatory Study

2.1 Scenario and Subjects

The study we are conducting takes place at *Vila União*, a suburb area in Campinas considered class D and E in terms of socio-economical indicators. A kind of "TeleCenter" hosts the activities; it is a physical space where several community initiatives for digital inclusion associated to the federal and local governments take place; *Casa Brasil, Centro de Referência da Juventude (CRJ)* and *Jovem.com* are some of those programs. *Casa Brasil* is a national initiative to promote digital inclusion. It is a public space where people have free access to a TeleCenter, library and basic computer courses. The same building also houses the *Jovem.com* and *CRJ* (Youth Reference Center). Both are projects from the city of Campinas and they all have the same objective: bring digital inclusion to the community as a means of opening the doors for social inclusion (Casa Brasil).

Within that community, we formed a group of 10 to 15 people representing part of the diversity we have in our population in its cultural, social and economical aspects. We named this group *Cenário Estrela* (or *Cenário**). Inspired by the numbers from the IBGE (Brazilian Institute of Geography and Statistic) we took percentages of the population in terms of gender, age range, literacy and income per family. We used these numbers to invite people from *Vila União* to take part of Cenário*. We counted on a partnership with the local Secretary for Citizenship from the City Hall to help us form this scenario. *Casa Brasil* is hosting the Cenário* and our team of researchers in regular practices twice a month. Figure 1 illustrates this place.

Figure 1 A view of *Casa Brasil*

On the first day of activities, we had 10 participants from the Cenário*. This group had 3 men and 7 women. We have attempted to have a group composition reflecting the less favored in terms of access to technology. For example, for ages around 30's the percentage of participants do not reflect the proportion we have in the country, as they are the part of the population who are most likely to have access to computer and internet, either at work or at college. Because our focus is to study the ones whose relationship with technology is lower or none we have chosen to maintain less people from the 30 to 49 age range. Nonetheless the point of view of the digitally literate still exists in the group and it is important when we think of the *Design for All* (two participants are well familiar with technologies).

The population of this Cenario* is very talkative and sincere, which significantly contributes for the success of our activities. They are able to express their thoughts and desires and are always willing to contribute and participate. Just as the Brazilian reality itself, the group is very heterogeneous: some of them had never touched a computer before while others are attending college. Despite de diversity, all members get along very well in a healthy and democratic atmosphere. Also, their commitment to the activities has contributed to the quality of our achievements.

2.2 Method and Practices

Before meeting the people of Cenário*, the group of researchers met to discuss and design the activities. The preparation stage is an important part of the process as it makes clear what the objectives are and how they should be achieved. In this preliminary work, all artifacts that are needed are designed and constructed and tasks are reviewed and distributed among researchers. Rehearsals may also be necessary; when you put yourself in the place of the user in the activities you are able to anticipate problems and also you may realize what other props might be needed throughout the practice.

The next step after the preparation is the activity execution itself. The first time with the users required a quick individual presentation and elucidation about the project and its purposes and goals. During the activity, all interactions were video taped and all conversations were recorded using MP3 devices. This has been done in accordance with the Brazilian rules of ethics in research. All the recordings were properly edited and saved to be used in the analysis, and are now part of a rich material that is of great importance for the research.

The analysis phase consisted of examining all material in order to understand the findings and derive preliminary conclusions regarding the objectives of that stage. The activities described in this article are part of an initial stage of design when we want to get acquainted with users' relationship with ICT. This accounts for the "gaps in user knowledge", an issue mentioned before, that is one of the three challenges mentioned by Shneiderman (2000) in gaining universal usability for Web-based services. Being aware of what users know is essential for development as it allows us to understand the users' context, which is a basic requirement of user-centered design (Minocha, 1999).

For this initial stage of the study we considered starting with the Story Telling technique (Muller, 2002) followed by a practice we designed to investigate the accomplishment of a common task (related to getting a second via of an id card) through different media. In the next sections we describe each activity and its preparation.

2.2.1 StoryTelling

The StoryTelling Workshop is a PD technique usually applied in the early stages of design in order to help the designers to identify and clarify the design problem (Rocha and Baranauskas, 2003). According to Muller (Muller, 2002), stories and storytelling in participatory work can function in at least three ways: 1. understand the product or service (when told by the end-user); 2. present what a designed service will do (when told by the designers) or 3. used as triggers for conversation. In our context, this method worked not only as a conversation startup, breaking the ice on the first day with the users but also helped us understanding the users' context.

With all participants gathered in a circle, each one told the group one case of success and one case of failure in using any type of information or communication technology. Simple examples of misuse of daily appliances like cell phones and digital alarm clocks were enough to make the group recognize common difficulties with technology in general. This identification added for the cohesion of the entire group. Knowing that all have abilities and difficulties made the group fell comfortable in sharing and freely participating in the activities without fear or constrain.

Activity preparation. The StoryTelling does not require much preparatory work. No extra materials besides the chairs arranged in a circle in the room and badges for the participants are necessary. The only action proposed beforehand, was that the researchers would think of simple cases of success and failure of their own, so that they would be able to share their experiences too, especially if the users would not feel comfortable to start with their stories.

2.2.2 Simulating a Service in Different Media

This activity aimed to investigate how comfortable the members of Cenário* feel getting information from different types of communication channel. An imaginary situation was proposed to the group: "*Paula, a married woman, 25 years old had lost her identification card and she does not know what she needs to order a new one*". One of the researchers played the role of Paula and she told the Cenário* about her problem. The group was supposed to help her and find out what were the necessary documents that she needed. For that, they were divided into four groups and each was taken into a

Figure 2 Interaction at S1 (left) and S2 (right).

different station: Station 1 (S1) an information booth simulation; Station 2 (S2) an Automated Response Unit simulation; Station 3 (S3) an iconic (graphical) computer screen simulation; and Station 4 (S4) a textual computer screen simulation. After talking to Paula at station S5 to present her with the answer, all groups got together again and shared their experiences. The stations are briefly described below.

S1 – Information booth simulation. Here the users interacted directly and personally with an attendant who followed a script to give the users all the information they would ask him for. This script was previously prepared and the desk had paper and pen in case they wanted to take notes.

S2 – Automated Response Unit (ARU) simulation. This station simulated an ARU: a call center service where the user interacts with a machine. ARU provides audible responses (pre-recorded sentences) to digital inquires from the user, usually by telephone.

The audible information on S2 came from a laptop computer that was kept hidden from the users behind a box. This box was built so that the audio source would be put in evidence, preventing the users from having any visual contact with the source of information.

Figure 3. Images (S3 above) and text (S3 below).

S3 – Iconic computer screen simulation. Cardboard pieces were make believe computer screens. With these artifacts the users were able to obtain the information from images and concrete objects displayed under a structure similar to the ones found in most of the internet sites. The researcher would hand in a cardboard at a time, accordingly to the users' choice, as if they were screens that came after users' mouse click.

S4 – Textual computer screen simulation. The same cardboard pieces were prepared for S4 as for S3, only that here there were no images, but only text. All the information was written and glued into the cardboards.

S5 – Paula´s place. This was the station where all groups went after they were done with their tasks. Paula waited here for them to return and tell her what documents she would need in order to apply for a new identification card. At the end of each report, Paula asked the users two new questions: whether the copies could be regular ones or the original documents were needed and if her husband, who had also lost his id card, would need the same documents.

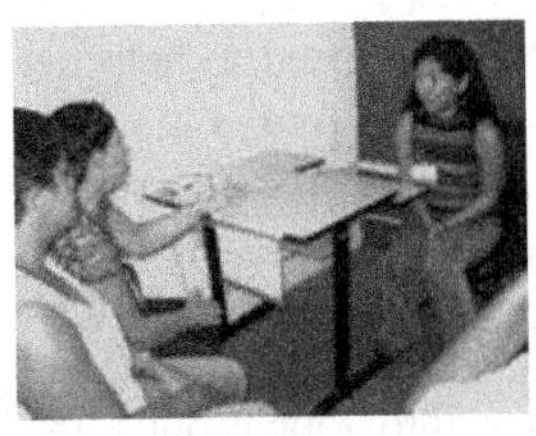

Figure 4. Results being delivered at S5.

Activity Preparation. Before going to *Casa Brasil* for the activities, all material was thought and prepared. For S1 there was the script for the help attendant, the cardboards with information for S3 and S4, and the pre-recorded sentences for S2. The content for all stations were basically the same: the list of documents necessary for acquiring different cards. This content is a simplification of what we found in our government Portal regarding this type of service.

Cardboards were chosen for this activity for the same reason most designers make use of paper prototyping: it is inexpensive and so familiar that users can feel more comfortable and not afraid of using (Rocha and Baranauskas, 2003), (Chand et al., 2006) them as they could be in front of a computer screen.

For S3, the cardboards were prepared as to offer as many images as possible and less written information. On the other hand, the screens simulations for S4 were text only. Both intended to deal with exactly the same information content.

The sentences recorded for S2 were saved on a laptop computer, but they actually had a manual process of activation. One of the researchers had to control the reproduction of the audio files accordingly to the users' response. Both laptop and researcher were hidden from users' sight through a box installed between them.

3. Preliminary Results

3.1 Warming up with Story Telling

A rich material resulted from this practice. All users had something to tell us and it all helped us to understand and identify abilities and barriers that stand between users and technology.

Four out of our 10 participants had an experience to share about the use of cell phones. Even knowing that the device has more features than just making and receiving calls, most people are limited to these functions. It is interesting to notice that cell phones are no longer restricted to the upper levels of the social-economic pyramid. According to IBGE (IBGE, 2005), 37% of our population (aged 10 years old or more) have a cell phone for personal use and most people have more than one unit, as per the National Telecommunications Agency (Anatel)`s data (Anatel, 2006), which shows that there are more than 120 million units supplied with mobile communication services. Mobile phones are also one of the preferred means of communication for deaf people in Brazil.

Other four participants reported their story of familiarity/unfamiliarity with computers. They all seem to understand that computers and internet access are a powerful source of knowledge but it seems that the elder consider these technologies something for the younger generation. Most people of age 50 or more think they do not have the skills to perform simple tasks in DVDs, mobiles or computers.

From this activity we can say that the StoryTelling workshop was an efficient mechanism of bringing people closer in the sense of knowing they all have something in common to share. Besides being able to recognize users familiarity with ICT, we were also capable of identifying vocabulary and metaphors used by Cenário*`s population. Their culture was also put in evidence, showing us a reality that we were not aware of. For example, one of the users mentioned that her mother did not allow her picture to be displayed on a social network website because it would reveal to strangers her bank account password. The concern about money was also noticed when another participant told us of the day when she did something wrong while using her telephone and she believes that a touch of a wrong button on the telephone made her loose all her telephone credits at once.

3.2 A Simple Task in Different Interaction Media

In average it took the groups about ten minutes to finish their tasks. The fastest, and by the way also the most effective one, was the group who worked at S1 with the help attendant. They got all the information in 3'38''. The group to spend the longest time interacting with his station was S2: they needed 21'06'' to deal with the sound device. The users who went to S1 were the only ones to bring to Paula the correct list of documents, although no group was able to correctly answer Paula's extra questions.

None of the groups took written notes of the information received; they all seemed very confident on the answer and had in memory the list of documents.

In S1, the group that interacted with a real person was able to maintain a regular conversation with the source of information. One of the members of this group mentioned to the attendant that Paula had her ID card stolen, which was immediately corrected by another participant. Paula's introductory talk was only oral.

Two of the three users on S2 left the station very confused and aware of the fact that they did not have the correct answer for Paula. The options from the audio simulation were too long for them to keep close attention. One of the participants from this group demonstrated very good memory and focusing capacities. She was able to specify details like how the picture for the new ID card should look like (front picture, recent and in plain background), and even named the documents that Paula should present in case she was single (married and single persons need different certificates)

It was interesting to note how a previous experience from a user that participated on S3 interfered in the task execution. She herself had had the same problem of losing her identification card. The difference was only that she is widow. By the time she experienced this situation though, she had trouble finding one of the necessary documents, which was her jobholder card[1]. One of the first screens presented to the users has a menu with three options. They are pictures of three different documents and the users were supposed to choose the one corresponding to the doc they had lost. She (with her group colleagues) promptly pointed to the jobholder card instead of the ID card and a sequence of different

[1] In Brazil, most employees have a document that looks like a passport, where all previous employers register the information about the job held by that person during the period that he or she worked for them.

screens followed. They only noticed the misjudgment when the list of required documents appeared with the lost ID card on it. The result that the group presented to Paula was not correct, and they even mentioned a document that does not exist: a single's certificate (meaning the person never got married. Actually she meant birth certificate, which does exist).

The text-only material from S4 worked smoothly and they were able to easily identify the right path through the screens. One of the participants was very attentive and read everything thoroughly. They did not take any notes though, and at the end they provided Paula with incomplete information.

3.3 Discussion and Lessons Learned

3.3.1 Users and their previous experiences

In different moments we could observe how users' prior experience can inadequately influence tasks performance.

Each life experience is unique and when we talk about a country with continental proportions like Brazil, we can count a variety of about 189 million different backgrounds. We can not prevent users from bringing their previous knowledge to the task, quite the contrary, but we must think of ways to make it in their benefit. Making the system simple in structure (Norman, 1990), minimizing distraction and interruptions is not enough.

3.3.2 Real life metaphor

The team to present the best result participated on a station where a real person informed the procedures to request a new ID card. In that situation, users' memory and attention were stimulated by visual aids (the attendant's complexion), audio (attendant's voice) and synesthetic (attendant's body expression and movement). That suggests us that redundant or complementary information are of great value when aiming for an interaction that resembles more like a human-human communication instead of a human-computer one (Reeves et al., 2004). As designers we must be careful though, not to overlap too much multimedia information as it could lead to a contrary effect (Reeves et al., 2004),(Leahy et al., 2003); it could even cause problems in memorization or learning (Jamet and Le Bohec, 2007).

One metaphorical embodiment of all this multimodal complexity could be that of an avatar. This idea came up from a conversation with a user who stated that the system should be like a "Geraldo online" (in reference to the name of the person who played the role of the attendant on S1). Indeed, this metaphor was taken in a prototype for a subsequent activity. The avatar would represent an attendant that gives all necessary information. According to Marcus (1998), the use of metaphors in interfaces can offer numerous advantages: the familiarity will require less training from users; it can add appeal; it may increase ease of learning; it can assist to a more direct communication; and cultural associations of user communities could be made. Still needs further investigation the implications of the anthropomorphic nature of the avatar for this population.

3.3.3 Multimodal interaction

The ARU simulation did not have a good acceptance in our experiment, but the use of auditory menus could help not only the visual impaired users but also the illiterate. Experiments on cell phones showed that such a menu is feasible (Eiriksdottir et al., 2006) and another study performed with blind children evidenced that sound at the interface can enhance memory and learning (Sánchez and Flores, 2004).

Graphic representations help users with subnormal vision, the illiterates, or even the hearing-impaired users. But alone, the images may not make sense or could be misleading, as we saw on S3, when sometimes more text was needed in order to complete the information that the picture provided.

Images and sounds are very helpful to add meaning to the context. They provide users with more resources for them to build they own reading strategies. This greatly facilitates understanding, but alone (i.e. only images or only sound) they are very less powerful.

3.3.4 The common ground

As heterogeneous as they can be, a group of users will always have something they share. It may be a point where they all agree or basic needs they all want to be fulfilled. There can be many different backgrounds that lead to different ways of thinking, but even so we can find more invariants than we could imagine. By focusing in these invariables we can move towards the Design for All. In the experiments we have conducted so far we saw that this recognition of the common ground is more easily noticed when we work directly with the user, in a participatory way.

3.3.5 Digital divides

From the StoryTelling activity we learned that even though they may not use ICT in its full capacity, the Cenário* is aware of the ICT' existence and its utility. Their concern relies mostly on the fear of loss, especially that of money. We perceived that during the StoryTelling activity, when they reported cases of losing credits due to misuse of the device or even a mystical relationship between having a picture shown on a web page and the bank account password being broadcasted. So, besides all initiatives for making technology available to the whole population, it is necessary to think of systems that would be so relevant that the contribution brought by its use would be greater than their fears or beliefs. Participatory practices contribute, may be in a small proportion, to the process of eliminating some of the myths by having the user actually use and see how the technology works and how they can benefit from it.

5. Conclusion

At the same time that ICT creates conditions for the emergence of knowledge societies, it can be a source of digital divide. In a developing country that has continental dimension and millions of inhabitants, the variety of its population is huge. The emergence of a knowledge society as suggested by UNESCO demands development for all. In this paper we address some issues regarding the challenges of making systems that can be accessed by everyone in these societies.

We presented how we are facing the challenge by describing the participatory approach we are using, illustrating it with the first practices in the Cenário*, with a group of users formed as to represent part of the diversity we have in the Brazilian's population, with its vast social, cultural and economical differences.

These activities were part of a project that aims at the development of inclusive social networks that make sense for people as a way of promoting their own life conditions by interacting in the digital world. The paper discusses the preliminary results and lessons learned from the activities designed for this study. The next steps include a more in deep study of the vocabulary mostly used by our target public and a test on the acceptance of some ideas taken from this present study, like the use of real life metaphors (avatars and voices) in informational web sites.

Acknowledgments. This project is funded by Microsoft Research – FAPESP Institute for IT Research. The authors thank all colleagues from NIED, InterHAD, Casa Brasil and IC Unicamp.

6. References

Anatel. http://www.anatel.gov.br (2006). Accessed 12 Feb 2008.

Casa Brasil. http://www.casabrasil.gov.br. Accessed 12 Feb 2008.

Chand, A. and Anind, K. D.: Jadoo – a paper user interface for users unfamiliar with computers. In CHI 2006, Montréal (2006).

Drucker, P.: Sociedade Pós-Capitalista. Pioneira, São Paulo. Translated from the original "The Post-Capitalist Society" (1993).

Eiriksdottir, E., Nees, M., Lindsay, J., Stanley, R. : User Preferences for Auditory Device-driven Menu Navigation. In Proceedings of the Human Factors and Ergonomics Society 50th Annual Meeting (2006).

IBGE (Instituto Brasileiro de Geografia e Estatística). http://www.ibge.gov.br/home/ (2005) Accessed 12 Feb 2008.

Information Society: the Next Steps. In Development Gateway Special Report. Available at http://topics.developmentgateway.org/special/informationsociety/index.do (2006) Accessed 12 Feb 2008.

Jamet, E. and Le Bohec, O.: The effect of redundant text in multimedia instruction. In Contemporary Education Psychology. Vol. 32, issue 4. Elsevier (2007).

Kensing, F. and Blomberg, J.: Participatory Design: Issues and Concerns. In CSCW, Springer, Neatherlands (2004).

Lanzara, G.F. The Design Process: Frames, Metaphors and Games. In Briefs, U. & Ciborra, C. & Schneider, L.: Systems Design For, With and By the User. North-Holland Publishing Company, Amsterdam (1983).

Leahy, W., Chandler, P. and Sweller, J.: When Auditory Presentations Should and Should not be a Component of Multimedia Instruction. In Applied Cognitive Psychology (2003).

Marcus, A.: Metaphor Design in User Interfaces. In Journal of Computer Documentation (1998).

Melo, A. M. and Baranauskas, M.C.C.: An Inclusive Approach to Cooperative Evaluation of Web user Interfaces. ICEIS, vol 1, p. 65 – 70 (2006).

Minocha, S.: Requirements Development in User-Centred System Design. In IEE Colloquium on Making User-Centred Design Work in Software Development (1999).

Muller, M.J.: Participatory Design: The Third Space in HCI. IBM Watson Research Center. Technical report (2002).

Norman, D. A.: The Design of Everyday Things. Currency and Doubleday, New York (1990).

Ramachandran, D., Kam, M., Chiu, J., Canny, J. and Frankel, J.: Social Dynamics of Early Stage Co-Design in Developing Regions. In Proceedings of the SIGCHI conference on Human factors in computing systems. California (2007).

Reeves, L.M., Reeves, L.M., Lai, J., Larson, J.A., Oviatt, S., Balaji, T.S., Buisine, S., Collings, P., Cohen, P., Kraal, B., Martin, J.C., McTear, M., Raman, T.V., Stanney, K.M., Su, H., Wang, Q.Y.: Guidelines for Multimodal User Interface Design. In Communications of the ACM. Vol 47, no 1 (2004)

Ribeiro, V. M.: Letramento no Brasil: reflexões a partir do INAF 2001. Global, São Paulo (2003).

Rocha, H.V. and Baranauskas, M.C.C.: Design e avaliação de Interfaces Humano-Computador. NIED/UNICAMP, São Paulo (2003).

Sánchez, J., Flores, H.: Memory Enhancement through Audio. In ASSETS'04. ACM, Georgia (2004).

SBC The Brazilian Computer Society (2006). Grand challenges in computer science. http://sistemas.sbc.org.br/ArquivosComunicacao/Desafios_ingles.pdf. Accessed 12 Feb 2008.

Shneiderman, B.: Universal Usability - pushing human-computer interaction research to empower every citizen. In: Communications of the ACM, May 2000, vol. 43, no. 5 (2000).

Unesco. Towards knowledge societies: UNESCO world report (2005). Available at: http://publishing.unesco.org/default.asp. Accessed 12 Feb 2008.

Varian, H. R.: Universal Access to Information. In: Communications of the ACM. October 2005, vol. 48, no 10 (2005).

World Bank, http://devdata.worldbank.org/external/CPProfile.asp?PTYPE=CP&CCODE=BRA (2006). Accessed 12 Feb 2008.

User Interface Input by Device Movement

Ryosuke Kokaji[1]**, Takako Nonaka**[2]**, and Tomohiro Hase**[3]

[1] Ryukoku University, Japan, t07m072@mail.ryukoku.ac.jp
[2] Research Center for Information Communication Systems, Japan, nonaka@rcics.hrc.ryukoku.ac.jp
[3] Ryukoku University, Japan, hase@rins.ryukoku.ac.jp

Abstract: This paper develops a user interface to input by moving a terminal device. First, a concept of the portable viewer with magnifying glass-like operation already proposed and remaining problems are described. Next, a new method to detect the device movement using optical flow is proposed. Then, a prototype with an embedded 32-bit MPU is built to verify the proposed method. As a result, basic functions necessary for the user interface operation of the image viewer were implemented in the proposed system.

Keywords: Embedded system, optical flow, portable viewer

1. Introduction

Recently, with portable information terminal devices becoming increasingly sophisticated, the number of various terminals with a small display as an output interface has increases. However, the amount of information that can be presented on the small display screen is limited, and therefore the terminal cannot display all information at once. The general methods for accessing hidden information are as follows: grouping and stratifying menus, and moving between the groups and hierarchies, and cutting out a part of a virtual screen, and scrolling to a hidden area. However, the user interfaces (UIs) share no common standards, and each model has different operation methods. This obviously can lead to confusion and frustration among the first time users. In addition, miniaturization and lightening the portable terminals are now important requirements, which reduce the area and volume occupied by special input interfaces on the device.

The authors have considered intuitive operation methods that are independent of previous usage experiences and knowledge about computers. The authors proposed a new UI whose input and output (I/O) relation is similar to a magnifying glass (Matsuda, 2007).

Please use the following format when citing this chapter:

Kokaji, R., Nonaka, T. and Hase, T., 2008, in IFIP International Federation for Information Processing, Volume 272; *Human-Computer Interaction Symposium*; Peter Forbrig, Fabio Paternò, Annelise Mark Pejtersen; (Boston: Springer), pp. 155–160.

This paper proposes a new method for detecting the device movement to achieve the proposed magnifying glass-like UI, and its validity is verified using a prototype.

2. Concept of the proposed UI

Figure 1 shows a system schematic to illustrate how the proposed magnifying glass-like UI works with the portable image viewer.

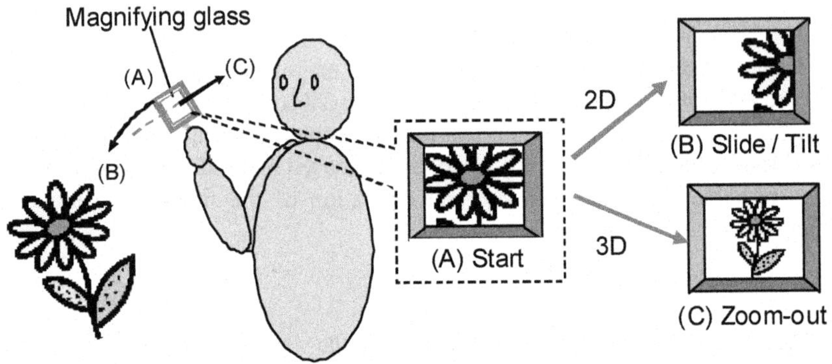

Figure 1 Concept of a portable image viewer with a magnifying glass-like UI

The proposed system aims at input operation without any pointing devices such as a mouse or a joystick. In addition, the relationship between the terminal movement and the output screen image on the small display is designed with the aim of achieving the relationship between the magnifying glass and an erect image formed by its convex lens. This will support those inexperienced with computer in understanding the operation method intuitively.

The required basic functions of the portable image viewer are as follows: rotation, movement, expansion and reduction of the image. It is therefore necessary to detect the gyration of the viewer body around three axes, as well as movement on three perpendicular spatial directions to achieve the proposed UI

3. New method of movement detection of the terminal device

There are many examples of detecting the terminal movement with acceleration sensors and gyroscopes. When detecting the terminal movement using these sensors, it is necessary to combine two or more sensors.

On the other hand, portable information terminal units equipped with small cameras normally, e.g. cellular phones, are increasing recently. Then, this paper devises the method of movement detection of the terminal using miniature cameras embedded in the terminal units efficiently.

Table 1. Characteristics of movement detection using acceleration sensors, gyroscopes, and miniature cameras.

Sensors & Devices	Displacement along an axis	Rotation about an axis	Characteristics & Problems
Acceleration sensor	Detectable	Undetectable *Detectable inclined angles by multiaxis	Integration error One sensor / axis
Gyroscope (Rate sensor)	Undetectable	Detectable	Integration error One sensor / axis
Miniature camera Optical flow	Detectable	Detectable	Processing load

When operating the proposed system, the user is expected to stand face-to-face with the output screen. The distance between the viewer and the user presumed to have 20/20 vision is limited to the range from 10 to 60 cm because of the screen size and resolution.

Therefore, a miniature camera is placed near and on the same surface as a small display. Next, optical flow is detected by using images captured by the miniature camera. Finally, the terminal's movement is estimated using the detected motion vector, and used to operate the UI. Figure 2 illustrates the hardware structure of the proposed viewer and its usage.

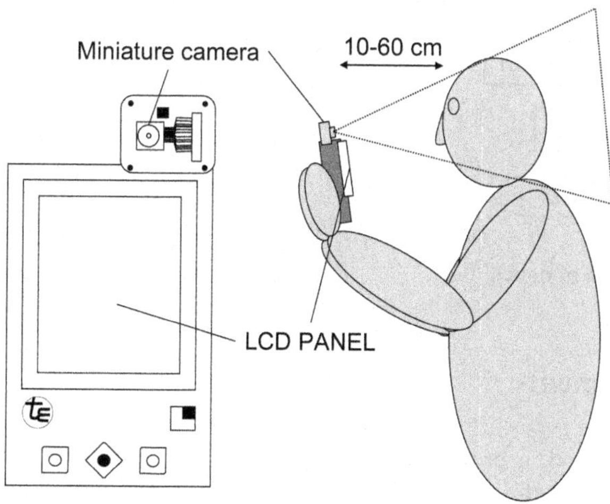

Figure 2 H/W structure of the proposed viewer.

When operating the proposed system, it is expected that the user's face will always be captured by the miniature camera located on the terminal. Thus, optical flow centered on the face is detected, and the movement of the device is presumed. Fig. 3 shows the detection algorithm of the optical flow (Matsuda, 2006).

The detected motion vectors show the movements of the surroundings as seen from the terminal, and the terminal's movements are then estimated from the relative motion.

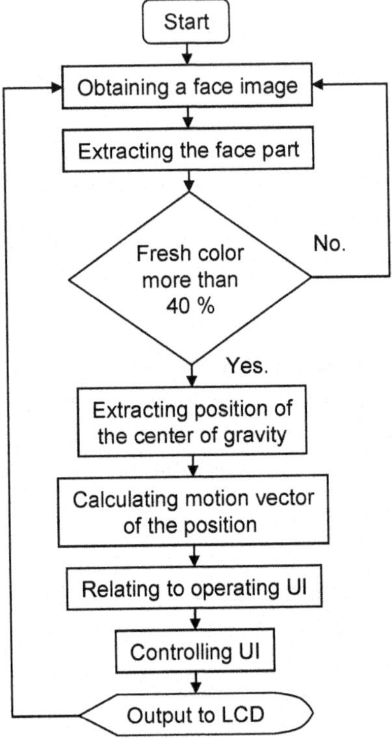

Figure 3 Detection algorithm of optical flow.

4. Verification Experiments

To verify the proposed method, a prototype with a miniature camera was built, based on a T-Engine embedded-device development standard platform. The prototype is shown in Figure 4 and the specifications are listed in Table 2.

The prototype operates on a 32-bit embedded MPU for consumer use, and its performance is reasonable for general portable terminals.

During the course of the verification experiments, the prototype was held and operated face-to-face by the user with both hands. The terminal was moved up and down and right and left in order to move the output images on the screen.

Table 2 Specifications of the prototype system

Item	Specifications
MPU	M32700 / 300 MHz
Memory	Flash 8 MB, SDRAM 32 MB
OS	Linux / M32R (kernel2.6.14)
LCD	320*240 (QVGA)
Camera	640*480 (VGA)

Figure 4 Photograph of the prototype.

As a result, when the terminal was moved up and down and/or right and left, the face movement was detected whenever user's face was captured by the camera, and the output image was then shifted using the motion vector.

5. Conclusions

Thus, the basic functions necessary for the UI operation of the image viewer were implemented in the proposed system. Furthermore, the desired magnifying glass-like operation was also achieved. It was possible to execute all computing and processing operations on a low specification system. Therefore, this proposed method could be easily implemented on a small portable terminal with a small display such as a cellular phone or a PDA.

References

Matsuda, M., Nonaka, T., Kokaji, R., Shimano, M., Hase, T.: Handy viewer with user interface according to device movement, Proc. ICCE 2007 on IEEE, pp. 319-320, Las Vegas (2007).

Matsuda, M., Nonaka, T., Hase, T.: An Input Method Using Hand Gestures for a Portable Remote Controller For Consumer Use, Proc. ICCE 2006 on IEEE, pp. 223-224, Las Vegas (2006).

An End User Development Model to Augment Usability of Rule Association Mining Systems

Elisa Albergaria, Fernando Mourão, Raquel Prates, Wagner Meira Jr.

Department of Computer Science, Federal University of Minas Gerais, Brazil
{elisa, fhmourao, rprates, meira}@dcc.ufmg.br

Abstract: One of the main challenges to a broader use of association rules data mining systems is their usability. In this paper we propose the End User Development Conceptual Model aimed at enabling the user to customize the interface of rule mining systems and create domain and problem specific queries. To do so, the user must be an expert user, both in the domain and system use (which usually requires knowledge of data mining technical concepts). The goal of the expert user is to create an abstract interface level that will allow a domain expert, with no knowledge of data mining, to use the system in specific problem situations. Thus, expert users can be perceived as co-designers of the system. An initial assessment of the model's usefulness and implementation feasibility was made.

Keywords: End-user development, semiotic engineering, data mining, rule association, customization.

1 Introduction

One of the outcomes of the Internet and the continuous growth of technology access is the creation of large volumes of data. The data mining field started over a decade ago (Webb, 2007) aiming at enabling people to extract useful knowledge from these large volumes of data. Different data mining techniques and methods that support knowledge discovery in distinct situations have been proposed.

Although many data mining (DM) systems have been developed and are in use, most of them still require the user to understand many aspects regarding the techniques or algorithms being used. Thus, one of the current challenges of the field is to decrease the complexity involved in using the systems, improving their usability (Kriegel, 2007) and broadening the use of such systems. In this paper, we present a model that enables at use time the creation of an abstract interface level that will allow domain experts who do not have data mining technical knowledge to make use and benefit from such systems. In order to provide a proof of concept for

Please use the following format when citing this chapter:

Albergaria, E., et al., 2008, in IFIP International Federation for Information Processing, Volume 272; *Human-Computer Interaction Symposium*; Peter Forbrig, Fabio Paternò, Annelise Mark Pejtersen; (Boston: Springer), pp. 161–174.

the model we focus on the usability challenges posed by one of the most popular frequent pattern mining techniques (Hipp, 2000), Association Rules (AR) mining systems.

In a nutshell, association rule (or just rule) mining generates sets of items in the form X → Y (Agrawal, 1994). Different interest measures are used to identify potentially relevant rules, such as confidence (conditional probability) or support (frequency of occurrence). For instance, a rule such as *Milk, Cereal → Coffee (80.00, 50.00)* means that 50% (support) of all purchases in this database include milk, cereal and coffee. It also means that, from all purchases that include milk and cereal, 80% (confidence) also include coffee.

Historically, DM systems have evolved from systems that supported just one step of the knowledge discovery process such as clustering or visualization (1st generation systems) to suites that support several steps of the process (e.g. WEKA (2008)), known as 2nd generation systems (Goldschmidt, 2005) (Piatetsky, 1999). Although these systems are currently very popular, they pose a great challenge to users. This challenge is due to their need for users to understand DM technical aspects, such as which is the best algorithm, the meaning of parameters, as well as their impact on the knowledge discovery. Particularly, in AR data mining systems the user must engage in a complex parameter setting (Kriegel, 2007) that has a direct impact on the quality of the knowledge discovered. Since these concepts are not part of users´ domain, learning them represents a high cost and challenge to users (Albergaria, 2006).

Recently, researchers have pointed out the need to create systems that are easier to use (Han, 2007) (Kriegel, 2007). Although there have been proposals in that direction, little has been done towards decreasing the necessary technical knowledge. The works that focus on the issue (Kirkland, 1999) have the disadvantage of limiting the data mining system to a specific problem. Our solution proposes a model that decreases the cost of learning technical concepts without limiting the power of AR mining systems.

In this paper we present the End User Development Conceptual Model (EDeM) that allows users to create domain specific solutions to be added to 2nd generation AR mining systems. In order to create customizations users must also have the required DM technical knowledge. These customizations will then allow users who do not have this technical knowledge to also make use of the system.

In section 2 we present other works aimed at improving usability of DM systems. Then we present theoretical framework that has grounded our work. Section 4 presents the proposed model (EDeM), its components and the preliminary evaluation conducted. Finally, we discuss this work´s contribution and next steps in the research.

2 Related Works

Some researchers of the data mining community have recently stated that one of their biggest challenges is to increase usability of DM systems (Kriegel, 2007). The main challenges in using AR mining systems involve having to learn DM technical concepts necessary to define parameters to execute the pattern search. This involves fine tuning an extensive number of parameters, interacting with a large volume of resulting rules, selecting relevant rules and interpreting them (Albergaria, 2006) (Hofmann, 2000) (Kriegel, 2007) (Mei, 2006). Efforts in dealing with one or more of these usability challenges have already been made.

Many authors have focused on supporting users in exploring large volumes of rules and identifying relevant rules by focusing on improving rule visualization or on reducing the number of rules. In an attempt to improve visualization different techniques and novel representations have been proposed (Han, 1996) (Rainsford, 2000) (Wong, 1999). To reduce the number of rules the main approaches have been either to generalize them by use of taxonomy (Domingues, 2005) or by pruning rules that are not interesting according to a specific measure (Srikant, 1997). Other works have focused on supporting users in understanding and interpreting a rule. In this direction visual metaphors (Hofmann, 2000) and semantic annotations have been proposed (Mei, 2006).

All these proposals are advances in improving the usability in rule mining systems. However, they still require users to learn DM technical concepts in order to interact with the system. Some works have as a goal supporting users in learning these concepts. In that direction, expert systems (4th generation systems) have been proposed to support users in decision making during the DM process (Goldschmidt, 2002). As users are guided to understand the process, they gradually learn the concepts.

There have been some works that have aimed at abstracting the required technical DM knowledge from users. To do so, they have created a DM solution that focuses on specific contexts (3rd generation systems). For instance, the Advanced-Detection System (ADS) detects fraud/violative behavior in the Nasdaq Stock Market according to NASD regulation (Kirkland, 1999). These systems are able to offer users DM systems that do not require the learning of technical concepts. However, they can only be used to solve a specific problem. Should the users′ problem evolve or change, or the user decide on a new approach, the system would no longer be useful (or a new version of the system would have to be developed).

In this paper we take an end user development approach to propose a solution that allows users themselves to create an abstract level to the system that could be used by other users without having to learn DM concepts. This solution allows for a new abstract interface level, without limiting the systems' applicability.

3 Theoretical Framework

One of the biggest challenges to develop usable systems is that requirements and contexts identified at design time often change at use time (Fischer, 2007). A proposed solution is to include End User Development (EUD) into the systems, that is, to allow users to adapt software to situational and novel uses that emerge at use time. EUD solutions vary from offering users opportunities to customize the systems all the way to including (re)programming components in the system (de Souza, 2005) (de Souza, 2006) (Fischer, 2004) (Fischer, 2007).

Fischer (2004) has defined meta-design as the design that allows users to become system's co-designers and has argued that this approach should be adopted by system developers as an EUD solution. In this paper we take this approach and propose the End User Development Conceptual Model (EDeM) that allows users to create domain specific solutions to be added to 2nd generation AR systems. This model is grounded on the semiotic engineering theory of HCI.

Semiotic engineering theory (de Souza, 2005) perceives an interface as a one-way meta-communication artifact in which the designer conveys to users who the system is aimed at, what problems they may solve with it and how to interact with the system to do so. As the user interacts with the system he understands the range of functions that can be performed, and range of contexts it may be used within. Semiotic engineering argues that the message should also present to him the rationale and design principles that have been followed in creating the system. In EUD systems this message should also communicate to users what parts of the system can be changed and how. As the users become co-designers they also become co-authors of the message being conveyed through the system.

4 End User Development Conceptual Model - EDeM

As we have pointed out, 2nd generation AR systems require users to learn technical concepts in order to engage in pattern finding activities (Albergaria, 2006) (Kriegel, 2007). Although some users are willing to learn them to be able to interact with these systems, others perceive it as just too high of a cost. Thus, the End User Development Conceptual Model (EDeM) presents an architecture model for an EUD module to be built and added to 2nd generation AR systems. The goal of this EUD module is to empower users to create extensions that allow for new interactions possibilities that do not require technical knowledge. To do so it offers users mechanisms to define context and problem specific mining tasks and create a new abstract interaction level in which to activate them directly.

The solution we have proposed was conceived based on a strategy pointed out to us in an interview with a 2nd generation system user that aimed at auditing Government expenses. He commented that he acted as the "miner" for his team, that is, he used the system to identify relevant patterns that could solve their problem, and then presented to his team what indicators they could use to identify expenses that should be audited.

Based on this strategy, EDeM considers two possible roles a user could take: expert or final user. The expert user (U_{Exp}) is an expert on the domain and understands the technical concepts required to interact with 2nd generation systems. Whereas the final user (U_F) may be a domain expert, but is not willing to endure the cost of learning all these concepts. In EDeM U_{Exp}s act as co-designers creating an abstract interface level for the U_F that is problem specific.

EDeM is composed of three components: the generator that allows the U_{Exp} to create the User Interface Abstract Language; the knowledge base that contains the U_{Exp}'s design rationale for the customization made; and the User Interface Abstract Language (UIAL) that is the resulting interface language with which the U_F will interact. Figure 1 depicts EDeM components and the communication among them. We next explain in detail each of these components and how they relate to each other.

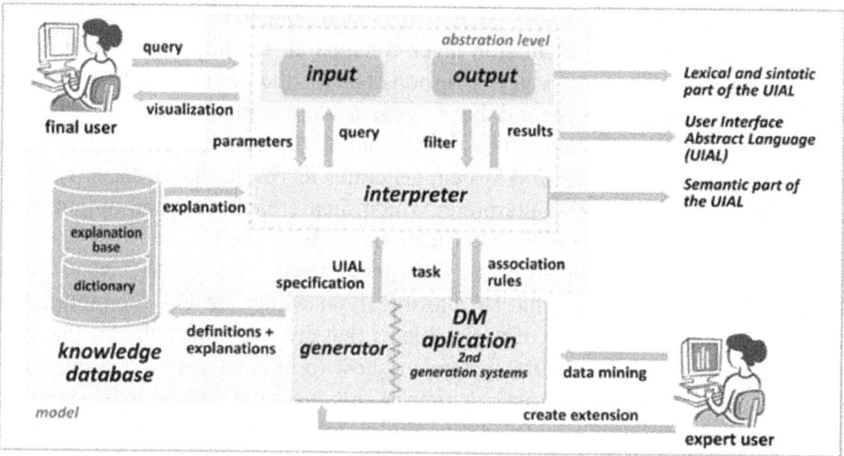

Figure 1 EDeM – End User Development Conceptual Model

4.1 User Interface Abstract Language

The User Interface Abstract Language (UIAL) is created by the U_{Exp} to the U_F. The UIAL comprises one or more problem specific queries that are an abstraction of a DM task. To create a query, U_{Exp}s define what parameters should be considered in the DM process, whether their values are fixed or will be defined by U_Fs. U_{Exp}s must also specify how the query, as well as it results, will be presented to U_Fs.

In order to quickly illustrate a potential UIAL, suppose that users worked for the agency responsible for auditing Government Purchases. U_F's responsibilities included identifying if any suppliers had been favored. Hence, the U_{Exp} could create a query as part of the UIAL such as: "*During the year of <DEFINE YEAR>, did any supplier win <OP [>]> <DEFINE PERCENTAGE>% of the bids to pur-*

chase product <DEFINE PRODUCT>?". In this query U_Fs would assign values to YEAR, PERCENTAGE and PRODUCT, which are all well known concepts in their domain. U_{Exp}s would also define the relation of these values to parameters (e.g. PERCENTAGE would be the value assigned to confidence), as well as set the values to other necessary parameters that are not mentioned at the UIAL level, such as support or AR algorithm selection. Using this query, a U_F could then ask the system "*During the year of **2008**, did any supplier win more than **50%** of the bids to purchase product **toner**?*". The results could be shown to him in a textual form, such "*ETA Inc. was the supplier in 55,19% of the bids to purchase toner.*"

Note that, as any language, the UIAL must have lexical, syntactic and semantic components. The lexical elements comprise the text of the query and of the results, as well as the parameters (and their values) that will be made available to U_Fs. The syntax is the possible combinations of the lexical elements at the UIAL. Finally, the semantic part is the behavior of the query, that is, the DM task that is executed once the query is activated.

In EDeM the UIAL is comprised by two subcomponents: the presentation layer and the interpreter. The presentation layer contains all the interface elements U_Fs will interact with. The interpreter is responsible for the communication between the UIAL´s presentation layer and the 2nd generation AR system. To do so it translates the specific query submitted by the U_F into a DM task that can be performed by the DM system. Once the DM system generates its result, that is the association rules, it is transmitted to the interpreter which then creates the final output to be presented to the U_F, based on the specifications made by the U_{Exp}.

In the semiotic engineering theory perspective, the UIAL is a meta-communication mainly from the U_{Exp} to the U_F in which the U_{Exp} communicates his views on what are (some) of the problems that are relevant to the U_F that may be solved by the underlying DM system and how to interact with the system. Notice that the U_{Exp} may be the main author of this message, but he is not the only one. The EUD module designers define the set of possible UIAL elements from which the U_{Exp} may be able to choose. Also, the designers of the 2nd generation DM system specify the set of possible mining tasks available to U_{Exp}s to associate to queries. Therefore, the UIAL is actually a message composed by the U_{Exp} and designers of the DM system and its EUD module.

4.2 Generator

The generator is the component in EDeM with which U_{Exp}s interact in order to create the UIAL (presentation layer and interpreter). The generator requires U_{Exp}s to define which elements will be presented to the U_F and how, from which it creates the presentation layer. The interpreter subcomponent is generated based on the definition of the DM task associated to each query at the UIAL.

In order to define the DM task associated to a query U_{Exp}s must decide which parameters are to be considered in the DM process, which ones have their values fixed at design time and which at use time. Although defining DM tasks involves mainly parameter setting the decisions involved in this setting may be complex (Albergaria, 2006) (Kriegel, 2007). EDeM does not intend to support U_{Exp}s in modeling a problem into a DM task, but rather, making this modeling available to other users who do not have the required technical knowledge to do the modeling themselves.

4.3 Knowledge Base

The purpose of the knowledge base is to allow U_{Exp}s to register their rationale for the abstraction they are creating. The knowledge base has two subcomponents: an explanation base and a dictionary. The explanation base stores all clarifications registered by U_{Exp}s regarding their decisions. The explanations are classified in two different levels, the ones that are to be made available to U_Fs, and the others that are more technical and intended to the U_{Exp} himself (or others U_{Exp}s).

For instance, associated to the query "*During the year of <DEFINE YEAR>, did any supplier win <OP [>]> <DEFINE PERCENTAGE>% of the bids to purchase product <DEFINE PRODUCT>?*" the U_{Exp} could associate the following explanation: "*This query allows you to explore the Government Purchase database and identify whether for a specific year and product any suppliers were favored. The underlying hypothesis is that no one supplier should win all the bids. In our experimentation we have found that those who win more than 40% of the times could already be considered candidates of having been favored. So the query considers the minimum PERCENTAGE value of 40%, but you could increase it if you would like.*". This explanation would allow U_Fs to understand the (intended) meaning of the query, as well as why there is a minimum value defined for PERCENTAGE. Thus, from a semiotic engineering perspective, this explanation is crucial to improve the quality of the U_{Exp} to U_F (designer to user) communication through the system.

An example of a more technical explanation would be to clarify why a specific fixed value has been chosen for a parameter (e.g. support = 0.27). For instance, "*This value was defined by testing different values in the Government Purchase Database and was considered a relevant value because ...* ". Notice that this explanation registers the rationale for a decision regarding the modeling of the problem into a DM task. Since support is not shown at the UIAL it is not intended to U_Fs but may be essential to document the *ad hoc* value chosen.

The other subcomponent of the knowledge base is the dictionary. The dictionary registers which elements of the underlying 2nd generation system interface language will be used in the UIAL, and how they will be translated to the UIAL. For instance, <PERCENTAGE> that appeared in the query shown above would have an entry in the dictionary such as the one depicted in Figure 2. The lexicon is defined by the U_{Exp} and contains its description in the UIAL (may have a variable, as well as the fixed text). The semantics is the element it represents in the 2nd generation interface. Finally, the U_{Exp} may also enter an explanation related to the element.

The dictionary has two main functions in the model. The first one is to lead the U_{Exp} to think about what elements of the DM system he believes should be part of the UIAL, as well as how they should be represented. The other is to support the U_{Exp} in maintaining the consistency among the elements that are shown in the input and output of UIAL, since these elements may be defined at distinct times in its creation process.

Lexicon: *<PERCENTAGE> % of the bids*
Semantics: *Confidence*
Explanation: *In this query confidence represents...*

Figure 2 Example of dictionary entrance.

Although the knowledge base component is not required to create or execute queries, it could have a major impact on how people use them. Semiotic engineering (de Souza, 2005) perceives it as an essential component. Inasmuch as it requires U_{Exp}s to add metalinguistic elements (i.e. elements that explain other elements/aspects) to the UIAL. These metalinguistic elements are a privileged part of the U_{Exp} to U_F communication, since they allow U_{Exp}s to send a direct message about their intentions or decisions. Also the dictionary acts in the model as an epistemic tool (i.e. a tool that increases the problem-solver′s understanding of the problem (de Souza, 2005)) to U_{Exp}s for it leads them to reflect about decisions being made regarding the UIAL.

5 Preliminary Evaluations

The solution proposed by EDeM was based on two main premises: (1) that expert users could use it to generate useful queries to final users who did not have DM technical knowledge; (2) that an EUD module based on its proposed architecture could be built for an existing 2nd generation AR system. Thus, our first assessment effort focused on collecting indicators about these two aspects: usefulness of the model and implementation feasibility.

5.1 Usefulness

Assessing potential usefulness of EDeM involved evaluating whether U_{Exp}s could actually create an abstract level that was domain dependent and relevant to U_Fs. An initial evaluation in that direction was done using scenarios (Carroll, 2000) and was carried out in two steps. The first one involved taking an existing DM task created in a 2nd generation DM system and defining queries that could potentially be useful to U_Fs. The next step was to verify whether other U_{Exp}s would be able to create abstractions that were relevant queries in different domains.

The first step was done by taking an existing problem that had been modeled as a DM task and creating an abstract level that described how a U_F would input the query and get the responses to it. The problem chosen was that of a real client of the DM research group at UFMG (Tamandua, 2008) the State Government Auditing Department, which is responsible for auditing public state agencies' bidding process and purchases. In 2006, rule mining tasks using a 2nd generation DM system, the Anteater (Guedes, 2006), had been defined in order to identify potential frauds in the purchasing database. The focus was on supplier favoring, (i.e. deciding in favor of a supplier by means not stated in the law). Originally, the mining task had been defined by experts of the DM group. Using this work as foundation, a scenario describing in detail how an abstract level based on the model could be created to give access to U_Fs to the issues they were interested in was created. (Examples used in the previous section are excerpts of this scenario.) The scenario showed how the model could support the creation of queries that would be interesting to intended U_Fs. Queries considered interesting were those that conveyed results described in the original final report.

The next step was to verify if other U_{Exp}s would be able to create relevant queries for different domains. This step of the evaluation was conducted as part of a class project for the DM course at the undergraduate level. In the project developed by the DM class (2nd semester, 2007), students had to model a real problem as a rule association mining task, and also create a model of the abstract level that would presented to a final user. In other words, define questions final users would be interested in asking, as well as possible domain specific responses that could be (automatically) generated from the resulting association rules. Thus, projects required students to interact with a real user to define real problems to be modeled, as well as to have access to the database containing the necessary information.

Projects were usually developed in groups of 2 students. Out of 12 projects that were handed in, 8 (assessed as having achieved initial goals) were considered for the model evaluation[1]. Projects were done in 3 different domains: quality of test questions on the University´s admission test (6), building electricity expenditure monitoring (1), and crime rates in the city (1).

All eight groups were able to create a successful abstract level (input and output). By successful we mean an abstraction that was problem specific and did not depend on understanding the underlying technical concepts. One group went beyond what was requested and actually implemented the query on top of Weka (2008) an open source DM system. In their project, students were also required to explain their proposed queries, as well as their modeling of the problem as a DM task. One interesting outcome was that even though the dictionary was not presented to them, most students included a mapping between representations used in the queries and elements of the DM system interface.

Both steps of the evaluation generated positive indicators that U_{Exp}s could create useful abstract level queries that would be relevant to U_Fs. Thus, the next step was to evaluate the feasibility of implementing a EUD module using EDeM that could be added to an existing 2nd generation DM system.

---

[1] After the semester was over students were asked if they authorized their work to be used for evaluating the model.

5.2 Implementation Feasibility

In order to evaluate whether an EUD module based on EDeM could be implemented involved actually implementing the model and analyzing the cost of adding it to a 2nd generation DM system. The system chosen was the Anteater (Guedes, 2006) (Tamandua, 2008), developed at Computer Science Department at UFMG, and to which architecture and code we had access to. Anteater is a DM platform that aims at providing scalable and efficient DM services (Guedes, 2006). The Anteater offers rule mining techniques to its users, but new frequent pattern finding techniques are also being developed to be added to the system. It has been used in projects with the Brazilian Government in different domains such as auditing, health and public safety.

The prototype to be added to the Anteater system was designed based on EDeM. Users have a choice of interfaces: query interface, extension creation interface, or original Anteater interface. In order to create the UIAL, U_{Exp}s must define the query to be shown to U_Fs; decide the values of parameters to be considered in the AR mining task; and specify aspects of the final presentation of the resulting association rules. The query input interface is always textual and the U_{Exp} defines the text to appear and which attributes (previously selected from the database) the U_F will be able to specify values for. When defining the query text to be presented, the U_{Exp} may also enter an explanation for it to be made available to the U_F. Figure 3 (A) and (B) show the screenshot of the extension creating interface for defining input query interface and its result at the UIAL, respectively.

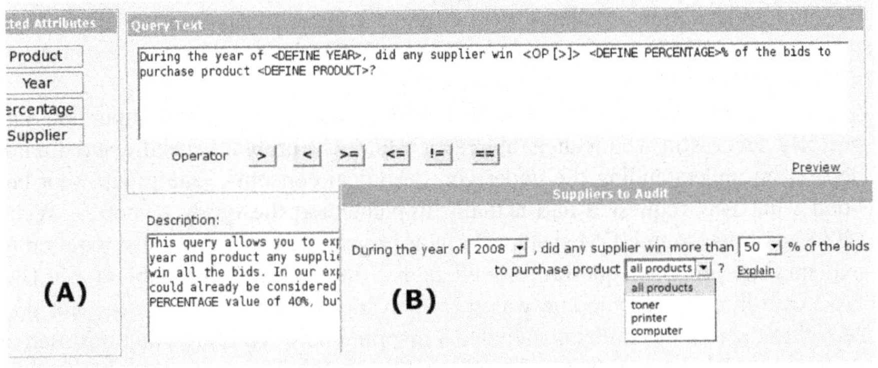

Figure 3 (A) Extension creating interface for defining query input; (B) UIAL input query interface.

The U_{Exp} also defines how the association rules should be translated to appear at the UIAL (Figure 4(A)) to the U_F (Figure 4(B)) and by which parameter it should be ordered (filters for attributes defined in the query are available to U_Fs).

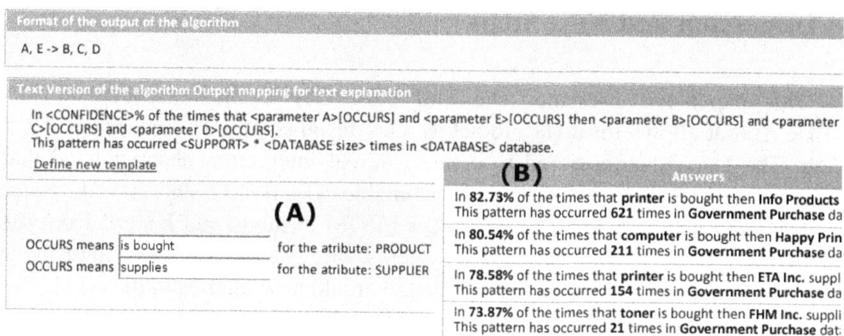

Figure 4 (A) Extension creating interface for defining query output; (B) UIAL results interface.

The prototype is being implemented (in Java and AJAX) as a proof of concept of EDeM to be added on to the Anteater system (implemented in Java and JSF). The Anteater has a distributed architecture based on major points of the knowledge discovery problem. Thus it contains 3 servers (database, data mining, and visualization) that contain a specific set of functionalities to be offered to users as services. All interaction during the processing of a user request happens over the Web, based on a user interface that controls the access to individual services. These services are in a 4th component, the application server.

EDeM required one more server to be implemented to contain its components, namely the generator, interpreter and the knowledge base. The interface module that controls the interaction with both expert and final users was added to Anteater´s application server. In this case, it made sense to keep together all the user-system interaction modules.

Even though the prototype is not fully functional yet, it has already shown that it is possible to create an EUD module to the Anteater. Most of the extension interfaces are already implemented and most of the UIAL generation works. Although the communication with the Anteater has not been completed yet, it has already been proven possible. The UIAL input query interface generates a list of parameters and their assigned values that are necessary to execute an AR mining task. Also the high level output is created from the association rule structure generated by Anteater (only textual form is implemented at this point, but table format and graphical presentations have been planned as alternatives). The main reason why the communication among the prototype and Anteater has not been completed is because rather than hard coding it into the system, we are investigating the possibility of using a markup language (namely, PMML) to transfer the information between them. This would make it easier to evaluate the prototype with other 2nd generation rule mining systems in the future.

6 Discussion and Next Steps

This paper has presented the End User Development Conceptual Model (EDeM) that allows for a U_{Exp} to act as a co-designer of a 2nd generation AR system. The U_{Exp} is empowered to create a novel interaction abstract level that is problem and context dependent and that enables U_Fs (who do not have knowledge of DM technical concepts) to interact with DM systems and benefit from them. Differently from 3rd generation DM systems, this solution does not limit the system´s applicability. U_{Exp}s can easily change or add new queries to the UIAL.

EDeM is founded on semiotic engineering theory which argues that designing is a communicative act. Thus, the model supports U_{Exp}s in creating their messages to U_Fs by providing them with a knowledge base (KB). The KB allows U_{Exp}s to register explanations and intentions regarding queries created, potentially improving quality of the U_{Exp} to U_F communication. It is also possible for U_{Exp}s to enter their rationale for the task modeling into the KB, supporting their reflection on values chosen, as well as documentation. In this direction, the dictionary also supports U_{Exp}s reflection on which elements of the underlying DM system to represent in the UIAL and how. Moreover, it helps maintain consistency of how DM system elements are represented in different steps of the UIAL creation process.

Different aspects of EUD systems may be analyzed to provide indicators about them. Fischer (2004) proposes that EUD languages should be considered according to their cost of learning to the user and scope. The EUD language proposed by EDeM has a low learning cost to U_{Exp}s (who already know the 2nd generation DM system´s interface language) and a low scope (since it is only useful for creating extensions for DM systems).

In that same direction de Souza and Barbosa (2006), in a semiotic analysis, show that EUD involve distinct signification systems for users and systems. Systems have a symbol manipulation perspective and should be considered in lexical, syntactic and semantic terms. Users have a communicative perspective and manipulate expressions, contents and intents. They point out that system and user perspectives do not have a one-to-one mapping. Thus, it is interesting to analyze the extension provided by EDeM according to these perspectives.

For the DM systems the extensions are meaning preserving, inasmuch as they can only manipulate lexical and syntactic levels (creating queries that combine parameters and their values). However, the systems cannot be semantically extended that is, no new functionalities may be added to them. In terms of U_{Exp}s communicative perspective the extensions are intent-preserving. As U_{Exp}s create a query they add a new expression to the system, associated to a specific content. Since the content and intent (what to achieve with the mining task) were already possible in DM systems, adding the queries has a rhetorical effect in the U_{Exp}s perspective. In analyzing the model, we must also consider its effect for U_Fs. They may not be directly involved in creating extensions but are the intended consumers for them. In their perspective extensions are full-fledged design that offers them new intent, content and expression, since prior to them they had no access to the system.

It is easy to see that creating extensions has a low cost for U_{Exp}s, but a very high benefit for U_Fs. As U_Fs understand the exploring possibilities of DM systems they may become willing to learn the technical concepts involved. In that direction EDeM may also be useful, since the explanations provided by the U_{Exp}s may support U_Fs in understanding the DM tasks associated to the queries, and gradually learning the technical concepts.

Although EDeM offers U_{Exp}s support to create the query and reflect about it, as well as register their rationale, it is not able to guarantee the quality of the extension created. The final extension may have a low quality for a number of reasons varying from poor choices of elements for the UIAL to a bad modeling of the domain problem into a mining task. This problem is not introduced by EDeM, since 2nd generation AR systems are also not able to provide any indicators on the quality DM tasks created by users.

The development of the prototype is currently being completed. Once it is fully functional and coupled to the Anteater, evaluation involving users (U_{Exp}s and U_Fs) will be performed. This assessment will provide us with indicators on the prototype itself, as well as on the underlying End User Development Conceptual Model. Although association rules mining is one of the most popular frequent pattern technique, there are a number of other relevant ones. Thus, we intend to investigate the model's applicability to other frequent pattern techniques and data mining methodologies. It would also be interesting to evaluate the cost of coupling the prototype to other 2nd generation AR mining systems, (e.g. Weka (2008)).

Acknowledgments The authors thank CAPES, CNPq, Fapemig and Finep for their support to their research and to the Anteater project.

References

Agrawal, R., and Srikant, R. Fast Algorithms for Mining Association Rules. In *Proc. of the 20th Int'l Conference on Very Large Databases*, Santiago, Chile, September (1994).

Albergaria, E., Prates, R., Almir, F., Rocha, L. and Meira Jr., W. Characterizing interaction challenges in data mining systems. *Procs of* IHC, 40-49, (2006) (*In Portuguese*).

Carroll, J. M. *Making Use: Scenario-Based Design of HCI.* MIT Press (2000).

de Souza, C. S. *The semiotic engineering of HCI.* MIT Press, Cambridge (2005).

de Souza, C. , Barbosa, S. A semiotic framing for end-user development. In: Lieberman; H., Paternò, F.; Wulf, V.. (Org.). *End User Development.* Springer, 401-426 (2006).

Domingues, M. and Rezende, S. Using taxonomies to facilitate the analysis of the association rules. *Procs of the 2nd Int'l Workshop on Knowledge Discovery and Ontologies*, 59-66 (2005).

Fischer, G., Giaccardi, E., Ye, Y., Sutcliffe, A. G., and Mehandjiev, N. Meta-Design: A Manifesto for End-User Development. *Communications of the ACM*, 47, 9, 33-37 (2004).

Fischer, G. Meta-Design: Expanding Boundaries and Redistributing Control in Design. *Proceedings of the Interact'2007 Conference*, Rio de Janeiro, Brazil, September, (2007).

Goldschmidt, R.; Passos, E.; Vellasco, M. An Action Plan Definition Assistant in KDD Process. *Procs of the 2nd International Conf. on Artificial Intelligence and Applications*, (2002).

Goldschmidt, R. *Data Mining – A Practical Guide*. Editora Campos (2005). *(In Portuguese)*

Guedes Neto, D., Meira Jr., W., Ferreira, R. A. C. Anteater: A Service-Oriented Architecture for High-Performance Data Mining. *IEEE Internet computing*, 10, 36-43 (2006).

Han, J., Y. Fu, W. Wang, J. Chiang, W. Gong, K. Koperski, D. Li, Y. Lu, A. Rajan, N. Stefanovic, B. Xia, and O. R. Zaïane. DBMiner: a system for mining knowledge in large relational databases. In *Proceedings of KDD*., 250-255 (1996).

Han, J., Cheng, H., Xin, D., and Yan, X. Frequent pattern mining: current status and future directions. *Data Min. Knowl. Discov*. 15, 1, 55-86 (2007).

Hipp, U Güntzer, G Nakhaeizadeh.. Algorithms for Association Rule Mining - A General Survey and Comparison. *ACM SIGKDD Explorations Newsletter*, 58-64 (2000).

Hofmann, H. Siebes, A. and Wilhelm, A. Visualizing Association Rules with Interactive Mosaic Plots, *Proc. Int'l Conf. Knowledge Discovery and Data Mining* (KDD). 227-235, (2000).

Kirkland, J.D.; Senator, T. E; Hayden, J. J.; Dybala, T.; Goldberg H. G.; Shyr, P.. "The NASD Regulation Advanced-Detection System," AI Magazine, 20, 1, pp. 55–67 (1999).

Kriegel, H., Borgwardt, K. M., Kröger, P., Pryakhin, A., Schubert, M., and Zimek, A. Future trends in data mining. *Data Min. Knowl. Discov*. 15, 1 (Aug.), 87-97 (2007).

Mei, Q.,Xin, D., Cheng, H. , Han, J., Zhai, C. Generating Semantic Annotations for Frequent Patterns with Context Analysis. *Proceedings of KDD*, 337-346 (2006).

Piatetsky-Shapiro, G. The data-mining industry coming of age. *IEEE IS*, 14, 6, 32–34 (1999).

Rainsford, C. and Roddick, J. Visualization of temporal interval association rules. In *Procs. of the 2nd. Int'l Conf. on Intelligent Data Engineering and Automated Learning*, 91-96 (2000).

Srikant, R. and Agrawal, R,. Mining generalized association rules. *Future Generation Computer Systems* 13 (2–3), 161–180 (1997).

Tamandua (Anteater) http://tamandua.speed.dcc.ufmg.br. Accessed in February (2008).

Webb, G. I. Editorial. *Data Min. Knowl. Discov*. 15, 1 (Aug. 2007), 1-2.

Weka,.http://www.cs.waikato.ac.nz/ml/weka/. Accessed in February (2008).

Wong, P. C., Whitney, P. and Thomas, J.. Visualizing Association Rules for Text Mining. In *Procs of IEEE Information Visualization*, IEEE CS Press (1999).

Investigating Entertainment and Learning in a Multi-User 3D Virtual Environment

Nicoletta Di Blas[1], **and Caterina Poggi**[2]

[1] Politecnico di Milano, Italy, nicoletta.diblas@polimi.it
[2] University of Wisconsin at Madison, USA, cpoggi@wisc.edu

Abstract: How can we include entertainment into an ITC-based educational experience? Edutainment is the blending of education and entertainment; it is about engaging, enjoyable experiences providing a learning value. Understanding better how to make a pedagogical intervention engaging, to stimulate significant effort and reasoning from all kinds of students, could generate a positive impact on education. We analyze the entertaining features in the design of Learning@Europe (www.learningateurope.net), an edutainment experience on European history for high-school students, based on a shared 3D virtual world. L@E has involved since 2004 over 6,130 students (aged 14 to 19) from 18 European countries plus USA, with highly rewarding results in terms of engagement and learning.

Keywords: Entertainment, Learning, Games, 3D Worlds, Virtual Reality

1. Introduction

How can we effectively include elements of entertainment into a technology-based educational experience, turning it into successful edutainment? Edutainment is the blending of education and entertainment; it is about experiences that are engaging and enjoyable, while at the same time providing a learning value. Edutainment can be seen as incorporating educational value into a leisure activity, so that players incidentally learn while having fun; or as making an educational experience more engaging, so that users are more motivated to learn. The focus on entertainment or education entails very different scenarios: in the first case users play in their spare time, to have fun: any learning resulting from it represents an additional value, which must not stifle the engagement; when learning is the primary goal, instead, the activity may take place in formal educational environments, e.g. schools, with a more "captive" audience, and it must provide substantial educational value; entertaining features may motivate users, enhance their experience, and engage also those who usually resist involvement.

Please use the following format when citing this chapter:

Di Blas, N. and Poggi, C., 2008, in IFIP International Federation for Information Processing, Volume 272; *Human-Computer Interaction Symposium*; Peter Forbrig, Fabio Paternò, Annelise Mark Pejtersen; (Boston: Springer), pp. 175–188.

This paper focuses on educational applications that are also fun to use. Gaining a better understanding of how to make a pedagogical intervention engaging, so as to stimulate significant effort, reasoning, and time on-task from all kinds of students, could generate a relevant - and highly needed - positive impact on education.
Many disciplines may contribute to the understanding of edutainment. In the first section we review relevant literature on game design, virtual reality for education, and motivationalist theories. Subsequent sections analyze the entertaining features in our edutainment programs based on shared 3D virtual worlds: we focus on Learning@Europe, an experience on European history for international high-school students which involved since 2004 over 6,130 students (aged 14 to 19) from 19 countries, with highly rewarding results both in terms of student engagement and educational benefits achieved. We describe the games and activities designed to make Learning@Europe engaging, and some evaluation results showing the elements that participants found most compelling. In the conclusions we summarize the critical elements that help creating engagement and learning in our programs and relate them to theory, drawing some general lessons.

2. Related work

This section presents theory and practice on the potential learning value of games.

Games are generating a growing interest in academic research as an emerging new media and design field. 70 years ago Huizinga (1938) recognized play as the root of many human activities: "Now in myth and ritual the great instinctive forces of civilized life have their origin: law and order, commerce and profit, craft and art, poetry, wisdom and science. All are rooted in the primeval soil of play".

First, we attempt to understand the nature of games.

2.1 A Taxonomy of Fun

What is a game? What makes games fun? Hunicke et al. (2004) proposed a taxonomy of fun, listing several game goals (Tab. 1). Caillois (1961) identified 4 major categories of games: AGON, based on competition (e.g. sports); ALEA, based on chance (e.g. gambling, dices); MIMESIS, based on role-playing of fantasy characters in imaginary settings; and ILINX, or dizziness, based on various states of mental frenzy (e.g. the roller-coaster). Bartle (1996) identified 4 reasons why people like playing MUDS (multi-user dungeons, or text-based online role-play games): *Achievers* like challenges and competition, *Explorers* are interested in the internal dynamics of the environment, *Socializers* take pleasure in social interaction, whereas *Killers* enjoy causing distress in other players. Koster (2004) identified several elements that produce pleasure, enjoyment, and possibly fun: stories, aesthetic appreciation, maneuvering to increase one's social status, flow, and visceral reactions. Finally, Prensky (2001) identified 11 reasons why games are engaging (shown in Tab. 1 or later in this article). Tab. 1 shows the overlapping of many elements identified by the various authors.

Table 1. A taxonomy of fun, according to different authors

Hunicke et al.	Bartle	Callois	Koster	Prensky
Sensation. Game as sense-pleasure.		(ILINX)	**Aesthetic appreciation**	Games are fun. That gives us **enjoyment and pleasure.**
Fantasy. Game as make-believe. **Narrative.** Game as drama.		**MIMESIS**	**Storytelling**	Games have **representation** and **story**. That gives us emotion.
Challenge. Game as obstacle course.	**Achievers** (Killers)	**AGON**	Fun = mastering a problem	Games have competition and **challenge**. That gives us adrenaline.
Fellowship. Game as social framework.	**Socializers**		**Social status** maneuvers	Games have **interaction.** That gives us **social groups.**
Discovery. Game as uncharted territory.	**Explorers**	(ILINX)	(Visceral reactions)	
Expression. Game as self-discovery.	(Killers)	**MIMESIS**		
Submission. Game as pastime			**Flow**	
		ALEA		

Good games are intrinsically engaging: motivationalist theories may help understanding them. According to Csikszentmihalyi (1991), for example, people work at their best when they are in a state of intense concentration that he calls *Flow*, with one or more of the following characteristics: clear goals (expectations and rules are discernible); high degree of concentration focused on a limited field of attention (a person engaged in the activity can focus and delve deeply into it); loss of self-consciousness: merging of action and awareness; distorted sense of time - one's subjective experience of time is altered; direct and immediate feedback (successes and failures in the course of the activity are apparent, so behavior can be adjusted as needed); balance between ability level and challenge (the activity is neither too easy nor too difficult); a sense of personal control over the situation or activity; and, the activity is intrinsically rewarding, so there is an effortlessness of action.

2.2 *Learning From Games*

Koster (2004) defines fun as “the feedback the brain gives us when we are absorbing patterns for learning purposes.” Gee (2003) identified 36 learning principles embedded in good video games (such as Situated Meaning, Multiple

Routes, Incremental Principle), which could be applied in educational settings. According to Prensky (2001), many of the elements that produce engagement in games could be employed also to make educational experiences more stimulating: interactivity, goals, outcomes and feedback, flow, win states, conflict-challenge, problem solving, social interaction, story. The affordances of games (action, structure, learning, creativity, social groups) and the affective states they produce (motivation, ego gratification, adrenaline, emotion) could be desirable features and outcomes also of a learning experience. Games can therefore be appropriate ways of achieving the entertainment goal in an edutainment experience : advocates of video games argue that games help developing abstract skills in probability, pattern recognition, and understanding causal relations (Johnson, S. 2005), have relevant learning principles embedded in their dynamics (Gee, 2003) and algorithmic understanding (Koster, 2004), offer opportunities to practice basic literacy skills (e.g. reading and writing) and learn valuable life lessons (Prensky, 2001). While the content of video games may be superficial, or even morally questionable, the underlying dynamics are extremely complex: game designers argue that players are not hooked by the game's "stage setting", but by its underlying logic (Koster, 2004). Through their compelling – when not addictive - reward structure, games keep players engaged for hours in tasks that few would find agreeable, while requiring from them a good degree of complex reasoning (Johnson, S., 2005). For example, the "probing cycle" (players probe the virtual world, form a hypothesis, test it by probing the virtual world again, and refine it basing on the outcome) is the basic procedure of the scientific method.

Experiments with introducing commercial video games in school curricula have been tried. Playing Civilization III in classroom environments taught students that history, geography and politics are all interrelated - while also providing insights on different students' motivations. For example minority students, usually uninterested in history classes which they see as propaganda, got involved in the game when they realized that it allowed them to change history, and play e.g. as Native American tribes who resisted European colonists and retained their lands (Squire, 2005). Here, strong identity issues are at play.

Exploring hypothetical history also helps understanding why events unfolded the way they did. Multiplayer historical role-play games such as Revolution (a collaboration of MIT, Microsoft and Colonial Williamsburg Foundation) can help students understand the interplay between personal or local concerns and the national, public concerns taught in history classes (Squire and Jenkins, 2003).

Games to teach literary analysis are also being explored. Prospero's Island is a single-player, nonlinear, open-ended game on Shakespeare's *Tempesta*, developed by the Royal Shakespeare Company and MIT to help students understand the play through an immersive experience "inside" it (Squire & Jenkins, 2003).

However, designing a multi-user game able to provide substantial learning as well as fun is a very hard task. An example is Arden, the World of William Shakespeare: while textually and historically accurate, full of Shakespearean quotes, characters and settings, it is not "gripping" as a game because "monsters were not part of the main game experience" (Castronova, 2007). The world failed to attract a sufficient number of players to allow using it for social science experiments, as originally intended. Online virtual worlds such as Second Life and role-play games with millions of players such as World of Warcraft are being used for testing social, politic, economic, even medical theories, e.g. the spreading of epidemic diseases (Johnson, 2007). However, such studies are hardly mainstream; there is also little research on how complex simulation games such as Sim City and Roller Coaster Tycoon could be used for learning (Squire, 2005).

Good video games are complex and very hard to beat: they require mastery of sophisticated worlds and often up to 50-100 hours to complete. Thus, they must be extremely good at teaching users how to play, through training modules, reward systems, and embedded dynamics encouraging exploration and experimentation. Games offer learning experiences that are embedded in action, experiential, multimodal; they involve discovery and self-knowledge; they situate meaning in embodied experiences and provide on-demand, just-in-time information (Gee, 2003). These are all desirable qualities for any educational intervention.

Given the enormous success of the videogames industry, with revenues larger than the GDP of a small country, a question emerges: is it possible to transfer the learning and motivational potential of entertainment into education (Barab et al., 2005), creating meaningful and engaging learning experiences?

2.3 Games, Learning and Virtual Reality

This section presents virtual reality applications involving elements of learning.

3D virtual environments hold great potential for creating experiences that are both engaging and educational. Yet this potential has just started being explored. Virtual reality has been used effectively in military or medical simulations of situations too dangerous, too expensive, or impossible to reproduce otherwise, e.g. for treating phobias or training firefighters. As for virtual environments used to teach curricular content, experiments include visualization and manipulation of complex, counter-intuitive concepts in scientific disciplines such as Physics (electromagnetic fields: Dede et al., 1997; visualization of molecular structures: Bergman et al., 2004; dynamics of flow lines: Bryson and Levit, 1992), Geometry (*Cyber-Math*: Naeve and Taxen, 2001) and Astronomy (Virtual Solar System Project: Hay et al., 2005). In some environments, users build rich, dynamic models of the phenomena studied, making theories explicit and developing a coherent understanding; in simulations they clarify misconceptions and construct a clearer understanding by manipulating variables and conducting systematic inquiry. 3D simulations have been used to build faithful virtual reproductions of ancient archaeological sites, such as ancient Olympia (Kenderdine, 2001); an ancient Greek house in Kassiopi (Mikropoulos and Strouboulis, 2004); or an architectural walkthrough of Monticello (Johnson, B., 2005). Second Life is being used by tourism and governmental agencies to attract interest on virtually-reproduced historical sites (such as Chichen Itza in Mexico); also, universities and cultural institutions use it to host online courses and public events. Finally, virtual experiences have been designed to teach scientific research methods through situated learning (*River City*: Dede et al., 2005), social and environmental issues

(*Quest Atlantis*: Barab et al., 2005; *Virtual Gorilla Exhibit*: Bowman et al., 1999), identity, personal and moral values (*Zora*: Bers, 2001). River City engaged over 1000 middle school students in observing problems, formulating and testing hypotheses, collecting data and deducing evidence-based conclusion while trying to solve the health problems of a 19th Century city (Dede et al., 2005). Quest Atlantis has been teaching thousands of elementary school children over the world about learning, playing and helping, encouraging them to take action in their local communities as part of the game (Barab et al., 2005).

This work examines another 3D-based edutainment experience, specifically addressing one of the most fascinating and challenging possibilities offered by the Web: cross-cultural interaction, the meeting of different cultural perspectives. In most educational virtual environments, content is mainly embedded in the 3D environment, and social interactions often play an accessory role. Little emphasis has been given to the potential educational value of cross-cultural interaction in virtual environments. The case study presented in this work, Learning@Europe (together with its "twin" Stori@Lombardia, their antecedent SEE - Shrine Educational Experience, and the new Learning@SocialSport) is to our knowledge the only example of a complex edutainment experience based on shared online 3D worlds that capitalizes on the different cultural backgrounds connected to the locations of participants, and uses their diversity as an asset to bring together and discuss multiple perspectives on a common cultural issue.

3. Designing an Edutainment Experience

Learning@Europe (L@E) targets high-school students aged between 14 and 19. Since 2004, it has involved over 6,130 students and teachers from 18 European countries and USA. L@E is a "blended" learning experience on European history, combining technology-based and "traditional" school activities. Students from four countries meet four times in a multi-user virtual environment, to play and learn under the guidance of two online tutors. Each synchronous session lasts about one hour. Between a session and the other, students interact with remote peers via online forums, study a set of interviews to renowned experts of history, and prepare research projects on their national identity and history, to compare with their remote peers. For details on the experience see (Di Blas & Poggi, 2007, Paolini & Di Blas, 2006).We highlight here the design elements crucial for engagement.

Interaction via chat with peers: during online sessions, students interact via chat both in the 3D world and in a parallel chat-only environment (dedicated to in-depth cultural discussion, under the guidance of a human online moderator).

Olympic Games: in Session 1 students play ability games requiring mastery of movement in the 3D world, e.g. flying through circles (fig. 2); their team partners help increase the score by answering cultural quizzes in the chat.

Treasure Hunt: in Session 2 students must search objects in a labyrinth, select those related to a given cultural clue, and ask a team-mate to confirm their choice.

Find your Way: in Session 3 two "blind" avatars must cross a path full of invisible obstacles, guided by their remote team-partners (who can see the obstacles). Correct answers to cultural questions in the chat facilitate the team mates' task (some obstacles become visible).

Questions via chat: the tutor in the 3D world asks quick factual questions on the contents showing visual aids (e.g. portraits of historical characters); the tutor in the chat-only environment asks conceptual questions requiring complex answers.

The main elements of fun in L@E, included for their motivational power, are the games (largely based on "physical' challenge, i.e. ability to control the avatar's movement), and the interaction with peers. The cultural questions, very important from an educational point of view, are also fun in that they are part of a "cultural" challenge: scores are awarded for the quickest correct answer. Tab. 2 relates these activities to the taxonomy of fun, also specifying their educational goal.

Table 2. Engaging activities in L@E according to the taxonomy of fun

Learning@Europe activities	Elements from Taxonomy of Fun	Educational functions
Interaction via chat with peers	*Fellowship*. Some moments during the sessions are especially dedicated to social interaction (e.g. presentations of students' classes and countries in the 1st session)	Increase of motivation. Building social ties as a basis for cross-cultural exchange. Practice of English as a second language
Olympic Games	*Challenge* (ability to control the avatar's movements, e.g. jumping, flying, steering; also, race against time) *Ilinx* (to some extent in the flying game)	Increase of motivation Development of skills for advanced interaction with 3D technology
Treasure Hunt	*Challenge*, "physical" and (mainly) intellectual. Avatars must find and select objects in a labyrinth basing on their knowledge of the contents. Also, race against time (and against the other team) *Discovery*: players explore a labyrinth *Alea*, in a small degree: finding objects in the labyrinth (and the right objects first) entails some luck. Yet, selecting correct objects by guessing is discouraged.	Increase of motivation Application and reinforcement of content-related knowledge Building of collaboration skills Development of skills for advanced interaction with 3D technology
Find your Way	*Challenge*, in terms of skilled control of the avatar through a path with invisible obstacles, and coordination with team partners for directions; race with time. An element of *Fellowship* is also present, as teams collaborate with each other	Increase of motivation Building of collaboration skills and team spirit Development of skills for advanced interaction with 3D technology
Questions via chat	*Challenge* in cultural terms: participants who first answer correctly gain scores. Time is relevant in factual questions; engagement is deeper in conceptual ones. *Storytelling* in some sense: discourse on history provides a "world" of characters, places, and events far beyond the physical and virtual spaces of participants	Motivating the study of contents Cross-cultural discussion and exchange of views on contents Reinforcing relevant concepts Testing knowledge of contents Detecting misunderstandings and knowledge gaps

A game must not necessarily include all the elements in the taxonomy in order to be entertaining; however, the presence of several elements can increase its appeal and extend it to a broader audience.

In the case of L@E, *Challenge* is present in both its "physical" and intellectual forms: controlling the avatar in the ability games and answering the cultural questions. This allows engaging both the "gamers" in the class (usually not the best students) and those more comfortable with "traditional" learning methods. Racing with time (playing or answering faster than others) increases engagement.

Fellowship is the other main entertaining aspect in L@E. Chatting with peers from different countries, playing in team with some and competing against others makes the experience more engaging. Interaction helps building social ties, critical for collaboration and cross-cultural exchange in cultural discussions, homework and team games. For non-native English speakers, social interaction is also a great incentive to English practice, an eye-opener on the importance of English, and a confidence-booster when students discover that they can understand each other.

The *Narrative, Storytelling* elements here are intended in a broader sense: while participants do not play any specific character, they discuss historical events which involve characters and stories, require some imagination, and provide a shared mental "world" beyond plain interaction with the 3D environment: while avatars seem just to be strolling around, students are actually constructing together an increasingly complex, detailed, and engaging picture of European history.

Sensation, Fantasy and *Mimesis* were not included in L@E because of flexibility and budget constraints: since each session involves a different set of countries, the virtual settings could not reproduce one specific place. Therefore the virtual environments resemble a content-neutral high-tech planet, with large round exhibit halls for discussions (Fig. 1) and ad-hoc environments for games (Fig. 2).

Figure 1. L@E: exhibit hall for presentations

Figure 2. Space for the Olympic Games

Two other 3D-based projects with different requirements did include *Mimesis* elements: in SEE, students began their discovery of the Dead Sea Scrolls from the museum where the manuscripts are preserved (Fig. 3); in Stori@Lombardia, on the Middle Ages in Italy, virtual settings resembled a medieval castle (Fig. 4); research assignments often involved role-playing dramatic historical situations (e.g. the council of a city under siege) and enacting them in the final session. This included elements of *Expression*: the same theme assigned to different groups never produced the same results. Teachers reported extremely high engagement.

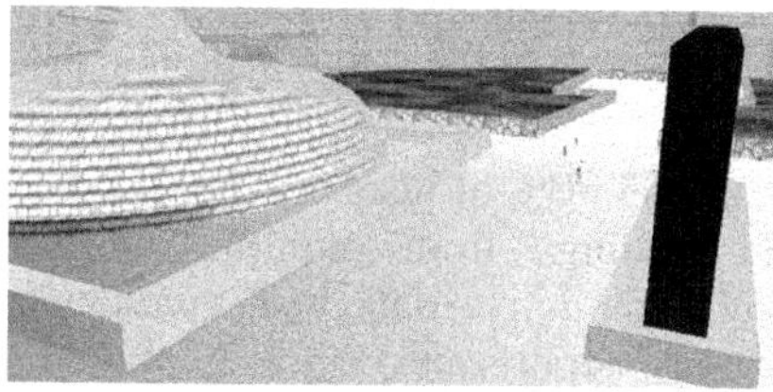

Figure 3. SEE: the virtual museum

Figure 4. The castle's dungeons in S@L

The design feature intended to generate *Flow* is the storyboard: the sequence of activities to take place during online meetings; it is carefully planned in order to never leave the students without something to do or a goal to achieve.

The *Alea* aspect has not been included in the design of any of the activities, except to a small degree in the Treasure Hunt, where participants may complete their hunt faster if the first objects they come across are the correct ones. Students who try answering cultural quizzes by guessing, however, are penalized.

A slight element of *Ilinx* might be present in the Olympic Game that requires flight (Fig. 2); however, movement in the 3D environment is not as life-like as in fly-training simulations. The fun factor here is mainly the "physical" challenge of controlling the avatar's movements, with an element of *Mimesis* in the excitement of a situation that has no counterpart in real life.

4. Investigating Fun Through Empirical Evidence

This section presents empirical evidence for the level of engagement achieved in the last two editions of Learning@Europe, investigating the reasons behind it as well as measuring the educational impact of the experience.

Learning@Europe involved between November 2004 and February 2008 over 6,130 students and more than 350 teachers from almost 190 schools in 19 countries. The other edutainment experiences based on a similar sequence of activities in shared 3D environments, i.e. Shrine Educational Experience (SEE), Stori@Lombardia (S@L) and Learning@SocialSports (L@SS), involved since 2002 another 3000 teenage participants from Italy, Belgium and Israel.

All experiences were monitored using similar sets of online surveys to teachers after every synchronous online meeting and to students before and after the experience. Online tutors wrote a report after every session. Chat transcripts, forum posts and students' works were collected and analyzed. A few on-field observations were conducted in some schools (at least 20 hours of class interaction have been video-taped). Focus groups with teachers were held in the early stages of each project and after deployment to assess the reliability of survey data.

While data from all projects through six years consistently show satisfactory results in terms of students' engagement and learning, we discuss here only the last two years of L@E, since other data are less comparable in terms of survey questions and rating scales. Approximately 1000 students and 100 teachers took part in L@E 2006-07, whereas L@E 2007-08 – the first experiment involving also an American school – involved 130 students and 8 teachers. Survey responses represent approximately 45-55% of students and 65-90% of teachers. To increase reliability, when possible we triangulated findings from multiple sources (teachers, students, online tutors), integrating quantitative data with qualitative evidence from open-ended survey questions, tutors' reports, etc.

4.1 Are L@E Activities Engaging?

Learning@Europe online activities involve classes of 15-25 students, of which only 4 at a time directly control a virtual user. Only 50-60% of survey respondents ever get to move an avatar in the 3D world or write in the chat, and less than 40% play a game: while they often take turns at the computers, in 70-80% of classes the most skilled are chosen to play the games. Most students spend the sessions grouped around their classmates, suggesting the next move or answer, with almost half of them never even touching the keyboard: yet, bored or slightly bored students are hardly ever more than 13%, while those rating an activity "really exciting!" (5 points out of 5) are around 20%.

In the surveys, 541 students in L@E 2006-07 and 57 students in L@E 2007-08 rated their engagement in L@E activities on a 5-points scale (1=boring, 3=quite interesting, 5=really exciting!). The highest average rating is for interaction with foreign peers (3.85 out of 5), immediately followed by the L@E experience as a whole (3.82): 30% of students found it "quite interesting" and 65% rated it "engaging" or "really exciting!". They appreciated the possibility to work in group with classmates (3.66), and enjoyed the questions via chat (3.58) even more than the games (3.5). Even "traditional" activities such as studying interviews (3.12) and preparing homework (3.19) excited some interest.

72 teachers over the two years also rated their students' engagement in online and class-based activities on a 5-points scale (1=not involved at all, 3=acceptable involvement, 5=enthusiastic involvement). Again, questions and discussions via chat (4.42 out of 5) score slightly higher than the games (4.36). Discussing the interviews (3.59) and doing the assignment (3.57) seem more engaging than studying the materials (3.47) and the forums (3.34), which few students used.

Asked to compare their students' engagement in L@E versus usual school activities, most teachers replied that it was equal and often superior in L@E: "*The involvement in the project looks really special and enthusiastic*", was the comment of a teacher. Others noted that the experience captured the interest of those usually not among the most involved: "*I could see that the more 'scholars' may not be the ones who are most at ease, on the contrary some find a field where to express themselves.*" Teachers, tutors and observers in schools agree that a sense of *Flow* was perceivable during most sessions. 78.9% of students in L@E 2006-07 (N=518) and 78.2% in L@E 2007-08 (N=55) declared they were so involved in the sessions that they lost track of time.

4.2 What Creates Engagement in L@E?

We now investigate the reasons behind engagement. Students were asked to select the session they liked best and give up to 3 reasons why they enjoyed it most (Fig. 5). The element selected by the largest number of students, "meeting new people", is *Fellowship* in the taxonomy of fun. While *Challenge* is the next preferred component, it seems that students enjoyed the overall competition, the cultural questions and the novelty of the experience more than the games.

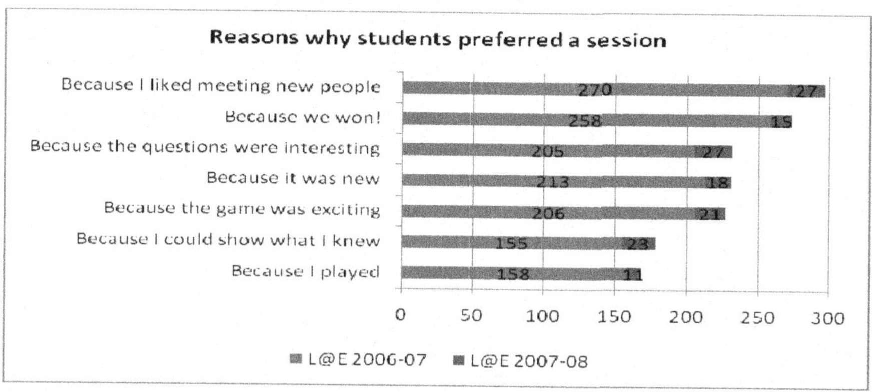

Figure 5. Preferences of students choosing max. 3 reasons why they enjoyed a session best. 539 respondents for L@E 2006-07 and 57 respondents for L@E 2007-08.

Students were also asked to rate the most attractive aspect of L@E on a 5-points scale (1=I didn't like it at all, 3=I liked it, 5=It was great!!). What they appreciated most was "doing something different from usual school activities" (4.06 out of 5, N=595): L@E appears as a welcome interruption of school routine. Immediately following came *Fellowship:* "meeting foreign students" (3.91) and then "studying history in a new way" (3.73). *Challenge* - "games and competition at school" - is rated the same as "studying with the computer": 3.58.

In conclusion, Learning@Europe is able to engage participants actively, even enthusiastically, through its novelty in school contexts, *Fellowship* and *Challenge.*

4.3 Is it Just Fun, or Are Students Also Learning?

We presents here also some evidence of Learning@Europe's educational impact. The students' engagement and fun were not achieved at the expense of their learning: on the contrary they have very likely enhanced it.

72 teachers over the two years rated the educational effectiveness of L@E activities on a 5-points scale (1=very poor, 3=good, 5=excellent). The most learning-effective activities are also the most engaging ones: the cultural challenge via chat (average rating: 3.76 out of 5) and the interaction and games in the 3D

world (3.74). "Proper" learning activities, namely preparing homework (3.70) and studying the interviews (3.66), were rated slightly lower; they involved respectively 40-50% and 70-78% of participants, but - unlike chat and games - could not be followed by the rest of the class through a projection screen.

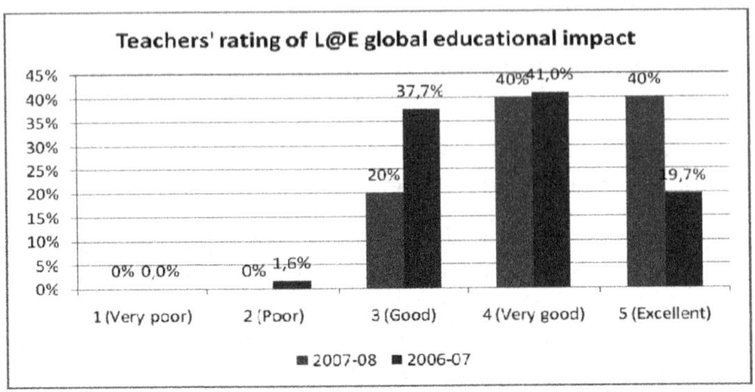

Figure 6. Ratings by 61 teachers from L@E 2006-07 and 5 teachers from L@E 2007-08

As shown in Fig. 6, almost the totality of teachers is satisfied with the educational impact of the L@E experience, and 60 to 80% are extremely satisfied. Teachers have different opinions on the educational effectiveness of individual activities: a class may benefit more from some and put less effort in others. Yet, all agree that the experience as a whole was beneficial in a variety of ways: students' improvements in terms of knowledge and understanding of history were rated high or very high by 35-50% of teachers, and good by 40-45%. Students practiced English as a second language, developed technological skills, skills for group work and new learning methods. The experience also changed their attitude towards foreign cultures, history, and school: 30-50% of teachers reported major or excellent improvements, and 40-50% good improvements. No teacher reported "no improvement at all" in any of the above areas.

5. Conclusions

Evaluation data from Learning@Europe (and related projects) shows that 3D world-based edutainment experiences can be educational and fun at the same time. The primary strategies to create engagement are *Fellowship* and *Challenge*, both in intellectual and "physical" form. Many elements in the experience design also favor *Flow* (Csikszentmihalyi, 1991).

Fellowship – interaction with remote peers and classmates – is an extremely powerful source of engagement for young users. Occasions should be created to encourage social interaction, which is also crucial in creating a solid basis for collaborative activities and cross-cultural exchange.

Challenge can be both “physical”, based on skilled control of the avatar, and “cultural”: quizzes and riddles, often even more engaging. In our programs, participants learn about the subject by reading a common set of materials before meetings: this puts all on equal footing for the cultural challenge; human online tutors acting as discussion moderators of and referees of games provide immediate feedback, adjusted to the quality of responses. Challenges must be adequate to the participants’ abilities: neither too easy, nor too hard. Cycles of evaluation and refinement over years allowed us to find a balance: games were redesigned based on participants’ feedback, ambiguous questions were revised, a simplified version of contents was created for younger students. Also, rules must be clear from the beginning and scores based on unambiguous criteria.

Educational effectiveness is based on the reading of quality educational material, providing a common ground for content-based interactions. Competition and interaction provide the motivation for studying; discussions with peers from different backgrounds make different perspectives emerge, challenge the interpretations of individual students, and force them to rethink critically at what they learned, reorganizing knowledge in a more complete picture. Assignments offer occasions for in-depth reasoning and research between online meetings.

Readings, assignments, online attendance at scheduled times are demanding tasks possible in the *“captive” situation* of formal school environments, where the alternative is another – often less attractive - mandatory activity: as a US students pointed out when asked what he expected from L@E, “*I'm not exactly sure, but it beats going through the same notions throughout an entire semester.*” The motivation of the teacher is extremely important: if teachers shows no interest in the program, students will hardly bother doing the required activities.

To favor a situation of *Flow*, the *sequence of activities* is defined in advance and the tutor is quick in setting a new goal once an activity is completed. This is especially important when collaborative time online is limited.

Educational benefits go far beyond the knowledge of the subject matter: collateral benefits, including development of technological, social, language and communication skills, ability to work in groups, discovery of new learning methods, attitude change, motivation and engagement, should also be considered: it is extremely unusual for a single educational activity to offer such a wide range of valuable educational benefits all at the same time. Opportunities for the professional development of instructors should also be considered.

References

Barab, S., Thomas, M., Dodge, T., Carteaux, R., and Tuzun, H.: Making learning fun: Quest Atlantis, a game without guns. Educational Technology R&D, 53(1), 86-107 (2005).

Bartle, R.: Hearts, Clubs, Diamonds, Spades: Players who suit MUDs. J. of MUD Research, 1(1) (1996). http://www.brandeis.edu/pubs/jove/HTML/v1/bartle.html. Accessed 11 Apr 2008.

Bergman, L., Richardson, J., Richardson, D., and Brooks, F.: View: an exploratory molecular visualization system with user-definable interaction sequences. Computer Graphics – Proc. SIGGRAPH 2004, 117-126 (2004).

Bers, M.U.: Identity Construction Environments: Developing Personal and Moral Values Through the Design of a Virtual City. J. of the Learning Sciences, 10(4), 365-415 (2001).
Bowman, D., Wineman, J., Hodges, L., and Allison, D.: The Educational Value of an Information-Rich Virtual Environment. Presence: Teleoperators and Virtual Environments, 8(7), 317-331 (1999).
Bryson, S. and Levit, C.: The Virtual Wind Tunnel. IEEE Computer Graphics and Applications, 12(4), 25–34 (1992).
Caillois, R.: Man, play and games. Free Press, New York (1961).
Castronova, E.: Two Releases: Arden I and Exodus. Terra Nova. 27 November 2007. Available at http://terranova.blogs.com/terra_nova/2007/11/two-releases-ar.html. Accessed 9 Apr 2008.
Csikszentmihalyi, M.: Flow: The Psychology of Optimal Experience. Harper Perennial, New York (1991).
Dede, C., Clarke, J., Ketelhut, D., Nelson, B., and Bowman, C.: Students' Motivation and Learning of Science in a Multi-User Virtual Environment. Presented at AERA 2005. Available:http://muve.gse.harvard.edu/muvees2003/documents/Dede_Games_Symposium_AERA_2005.pdf. Accessed 10 Feb 2008.
Di Blas, N., and Poggi, C.: *European Virtual Classrooms: how to build effective 'virtual' educational experiences*. Virtual Reality: Special Issue on Virtual Reality in the e-Society. Springer, London. http://www.springerlink.com/content/r264167mt8654q81 Accessed Feb 08
Gee, J.P.: *What Video Games Have to Teach Us About Learning and Literacy*. Palgrave, New York (2003).
Hay, K.E., Kim, B., and Roy, T.C.: Design-Based Research: More than Formative Assessment? An Account on the Virtual Solar System Project. Educational Technology, 45(1), 34-41 (2005).
Huizinga, J.: Homo Ludens: A Study of the Play Element in Culture. Beacon Press, Boston (1938)
Hunicke, R., LeBlanc, M., and Zubek, R.: MDA: A Formal Approach to Game Design and Game Research. Proc. of the Challenges in Game AI Workshop, 19th National Conference on Artificial Intelligence (2004).
Johnson, B.: Place-Based Storytelling Tools: A New Look at Monticello. Proc. Museums and the Web 2005. Archives and Museums Informatics, Toronto (2005).
Johnson, C.Y. Online gamers become guinea pigs. Epidemics uncorked in virtual worlds. The Boston Globe, 25 Aug. 2007. Available: http://www.boston.com/business/technology/articles/2007/08/25/online_gamers_become_guinea_pigs/. Accessed 11 Apr 2008.
Johnson, S.: Everything Bad is Good for You. How Today's Popular Culture is Actually Making Us Smarter. Riverhead Books, New York (2005).
Kenderdine, S.: 1000 Years of the Olympic Games: Treasures of Ancient Greece. Proc. Museums and the Web 2001. Archives and Museum Informatics, Pittsburgh (2001).
Koster, R.: A Theory of Fun for Game Design. Paraglyph Press, Scottsdale (2004).
Mikropoulos, T. A., and Strouboulis, V.: Factors that influence presence in educational virtual environments. Cyberpsychology and Behavior, 7(5), 582-591 (2004).
Naeve, A., and Taxen, G.: CyberMath: Exploring Open Issues in VR-Based Education. Proc. SIGGRAPH 2001, Educators Program, 49-51 (2001).
Paolini, P., and Di Blas, N.: Multi-User Virtual Environments for Education: A European Experience. Proc. E-Learn 2006, AACE Press, 1383-1394 (2006).
Prensky, M.: Digital Game-Based Learning. McGraw-Hill, New York (2001).
Squire, K.: Resuscitating research in educational technology: Using game-based learning research as a lens for looking at design-based research. Educational Technology, 45(1), 8-14 (2005).
Squire, K., and Jenkins, H.: Harnessing the power of games in education. InSight, Vol.3, 5-33 (2003). http://website.education.wisc.edu/kdsquire/manuscripts/insight.pdf. Accessed 11 Apr 2008.

Openphone User Engagement And Requirements Solicitation in Low Literacy Users

T. Jama Ndwe[1], Etienne Barnard[2], Mqhele Dlodlo[3], Daniel Mashao[4] Christiaan Kuun[5], Aditi Sharma[6]

[1] CSIR, South Africa, JNdwe@csir.co.za
[2] CSIR, South Africa, EBarnard@csir.co.za
[3] University of Cape Town, South Africa, Mqhele.Dlodlo@uct.ac.za
[4] SITA, South Africa, Daniel.Mashao@sita.co.za
[5] CSIR, South Africa, CKuun@csir.co.za
[6] CSIR, South Africa, ASharma1@csir.co.za

Abstract: The OpenPhone project aims to design an Interactive Voice Response (IVR) health information system that enables people who are caregivers for HIV/AIDS infected children to access relevant information by using a telephone in their native language of Setswana in Botswana. The system lowers accessibility barriers since it is accessible to illiterate users and the community of the blind. The design utilizes usability engineering methodology in order to ascertain that the end product is usable, efficient, effective and satisfactory to the targeted users who are predominantly females, ranging from semi-literate to illiterate adults but nevertheless numerically literate. The paper describes the methodologies that were used to obtain information from the target user population. Based on the information gathered, we are now able to begin the initial design of the OpenPhone system.

Keywords: Usability engineering, User Requirements, Participatory Design, OpenPhone, Botswana-Baylor Children's Clinical Centre of Excellence

1. Introduction

In order to develop a system that meets the users' anticipation of the system the developers have to depend on the information that is provided by the users or anticipated users of the system (Lynch and Gregor, 2004). In this paper the development of collaboration between the researchers and the anticipated target users of the OpenPhone system is presented. Potential users were requested and encouraged to participate in the design process as a strategy to ensure that the product designed meets their needs and is usable to them. User engagement is particularly important for this targeted user group because of a cultural aspect, "where a questioner invariably gets positive answers as a matter of politeness even

Please use the following format when citing this chapter:

Ndwe, T.J., et al., 2008, in IFIP International Federation for Information Processing, Volume 272; *Human-Computer Interaction Symposium*; Peter Forbrig, Fabio Paternò, Annelise Mark Pejtersen; (Boston: Springer), pp. 189–193.

if these are not actually true! Criticism is seen as a sign of disrespect" (Blake and Tucker, 2006).

2. Caregiver focus groups

The focus group approach was chosen as the primary methodology for acquiring initial user needs from the intended target users of the OpenPhone system. This methodology is pragmatically appropriate for this particular user group because focus groups:

- *Do not discriminate against people who cannot read or write*
- *Can encourage participation from those who are reluctant to be interviewed on their own (such as those intimidated by the formality and isolation of a one to one interview)*
- *Can encourage contributions from people who feel they have nothing to say* (Kritzinger, 1995).

These characteristics about focus groups and the use thereof fit the intended user population well in the milieu of the unique situation and challenges of this user population as discussed in section 1 and in the abstract.

The caregiver focus group meetings had 3 primary objectives.

- The first was to study the user characteristics in order to compile and develop a user profile. The first stage in the usability process is to study the product's intended users (Nielsen, 1992).
- The second was to allow the targeted users to voice their opinions as to what concerns would they like the proposed system to address.
- The third objective was to engage the targeted users to form a coalition in the design of the system through participatory design which intends to involve the targeted users in all phases of the project. This particular focus group engagement was focused on introducing the system to the target user population and gathering opinions, beliefs, and attitudes about issues of interest about the proposed system.

There were 11 participants on the first day and 16 on the second and were all female with only one male participant on the second day. Both sessions started with a welcome speech that also thanked the participants for attending the focus groups. All conversations and interactions with the caregivers were conducted in Setswana with the aid of 2 moderators who are both fluent in English and the local language of Setswana. One of the moderators was a local Botswana citizen and resident which has helped the research team in comprehending the local cultural nuances which would not have been understood by any other means, not even by the other moderator who is fluent in Setswana but not a resident nor a citizen of Botswana.

The moderators made it clear to the participants that the research needs to learn from them as to what was needed to be addressed by the proposed system. The moderators informed the participants that the system to be built would only succeed if the participants, who will be the users of the system, collaborated and partnered with the research team in building the system and that the team was respectfully asking for their cooperation. The research team is fully aware of the limitations of how much the participants can contribute to the design of the system, but nevertheless, the research team regards the participants, who are future end-users of the system, as experts developing and defining tools for their own use within their own environment (Schuler and Namioka, 1993).

The participants then enthusiastically engaged in the discussion and brainstorming of concerns that they felt the OpenPhone system should address. The discussion was again conducted by the 2 moderators and 2 observers were taking notes of the discussion. Both focus group sessions were recorded on a computer with a microphone connected to it. Their enthusiasm was evidenced by the fact that they would chat amongst each other and discuss amongst themselves what they think is necessary to be available on the proposed system before giving their views to the moderators. Naturally some participants were more talkative than others but the moderators encouraged those who were less talkative by engaging them in the discussion and asking them what they consider important and should be made available in the proposed system. The focus groups took an average of 105 minutes each and at the end of the focus groups the participants were then thanked for their participation.

3. Summary of findings

Unexpected issues on social services such as government grants were brought up by the participants but unfortunately these issues cannot be addressed directly by the system. Issues that were contemplated by the designers as of high importance such as caregiver psychological support were perceived as of low priority by the participants. When the participants were probed about this issue they stated that they get psychological support through strong immediate and extended family support. This support can also be communal which is typical of the Tswana culture whereby family and close community members are supportive towards other community members especially in the rural areas where those communal values are still maintained.

4. Benefits of the focus group meetings

Conducting the focus group meetings has had an impact on the designers' beliefs in terms of testing the designers' general assumptions whereby some of the information that the designers imagined as important to the caregivers was not viewed as such by the participants. This supports Robinson's notion of difficulty in anticipating a system's use in its actual applied work environment (Robinson, 1993). The meetings have enabled the designers to gain the targeted users' inputs on what their information requirements are and to eliminate unnecessary elements that the users don't need in the proposed system, which they have other ways to cope with. The meetings have also helped the design team in building a persona which is a model user that the design team creates to help understand the objectives, needs, and behaviours of the target users who will use the system interface. Benefits of creating an OpenPhone persona are:

- Creating a persona has assisted the designers to approach the design more objectively, with their target user in mind, instead of their own views and beliefs. Instead of asking, "How would I use this system?" the designers are now asking, "How would the caregivers use the system?"
- In using the persona as the target character, the designers are more capable of identifying how the caregivers will interact with the design. This enables the designers to gain an insight about the design and system usage that they wouldn't have gained in any other way.
- Puts all the design team members on the same page as far as to who the design is intended.
- It enables designers to put themselves in the shoes of the target users.

In sum the persona helps the researching designers make a smooth transition between user requirements and the design, which will benefit the overall design of the system. The created persona is a primary persona and is expected to evolve as the designers gain more knowledge about the targeted users of the technology.

5. Conclusions

User needs gathering is a way of animating and furnishing influential information into the design process that will have impact in the manner in which the system is designed. In conducting real user observations the researchers' findings get to be based on realities, not preconceptions. Users bring about things that the researchers would erroneously consider unimportant in a focus group. On the other hand users also remark on the things that the researchers thought to be essential but users don't see to be beneficial to them as they have other ways of dealing with such issues beyond the capabilities of the system.

The contacts made with the actual prospective users enable strong relations between the users and the researchers to be formed. These relationships are envisaged to create trust and understanding between the parties in order to devise a bond with common goals of designing a truly usable system.

In a new and modern design the designers may be mislead into assuming that there is no need for user studies because the product idea is new and ground-breaking to the target users and therefore there is no useful information that can be provided by the users. On the contrary, it is essential to observe and interview people in order to understand how they cope in doing things the traditional way before bringing in the new way of doing things. Through the interactions with target users the researchers may discover that they are solving the wrong problem, or that they have overlooked some other more important problems that need to be solved. Researchers may also find that there are features of the old way of doing things that work well which need to be reserved and incorporated in the new design. For example, in the OpenPhone system there is a need to use the same language and terminology that is usually used by the lecturing staff during the lectures at the Botswana-Baylor Children's Clinical Centre of Excellence as the users are accustomed to those terms and language and not the scientific terms and language as acquired by researchers from formal literature. The Botswana-Baylor Children's Clinical Centre of Excellence is the clinic where the caregivers normally go to get information on giving care to HIV infected children, where the focus group meetings were held.

We have found out that although the participants lack knowledge about technical matters on how to build an appropriate IVR system, they are rich in common sense knowledge about their needs and their typical concerns on care-giving issues. Both scientific and common sense knowledge is important in formulating a holistic solution.

Based on these focus group meetings and the persona that has been created the designers are able to turn the requirements information into functional specifications before beginning the initial design of the system.

References

Blake, H. and Tucker, D. 2006. User interfaces for communication bridges across digital divide. *AI Soc.* 20, 2 (Feb. 2006), 232-242.

Kritzinger, J. (1995) Qualitative Research: introducing focus groups. *BMJ* 311:299–302.

Lynch, T. and Gregor, S. 2004. User participation in decision support systems development: influencing system outcomes. *Eur. J. Inf. Syst.* 13, 4 (Dec. 2004), 286-301.

Nielsen, J., "The usability engineering life cycle," *Computer* , vol.25, no.3, pp.12-22, Mar 1992.

Robinson, M. 1993. Design for unanticipated use.. In *Proceedings of the Third Conference on European Conference on Computer-Supported Cooperative Work* (Milan, Italy, September 13 - 17, 1993). G. de Michelis, C. Simone, and K. Schmidt, Eds. ECSCW. Kluwer Academic Publishers, Norwell, MA, 187-202.

Schuler, D. and Namioka, A. (1993). Participatory design: Principles and practices. Hillsdale, NJ: Erlbaum.

Complex and Dynamic Data Representation by Sonification

Maher CHEMSEDDINE[1] **and Monique NOIRHOMME-FRAITURE**[2]

[1] FUNDP Institut d'Informatique, Belguim, maher.chemseddine@fundp.ac.be
[2] FUNDP Institut d'Informatique, Belguim *monique.noirhomme@fundp.ac.be*

Abstract. So far, data representation has been based on visuals. The huge size of data verges on overuse of the visual capability. Thus, there is a need to reduce the almost exclusive use of visual techniques to represent data in order to increase our perception bandwidth. In this research, we aim to solve this problem by integrating the audio component. More precisely, we are interested in representing data using musical melodies. This paper presents a model for music elements based on human emotion, to express alert messages displayed by computer network monitoring.

Keywords: data representation, sonification, emotional music.

1. Introduction:

In dynamic and complex datasets, we have many parameters represented or modeled which make it difficult to distinguish those using only visual methods. The capability of looking at some dimensions while listening to others allows one to process more information at once and make better correlations. Also, auditory accompaniment clearly leads to an improved visual information reception.

Sonification (Herman, 2006) exploits our auditory channel perception. In contrast with the vocal interfaces, sonification uses no-speech audio to convey information. There are various methods of sonification, such as Auditory icons (Conversy, 2000), Earcons (André, 1993), Audification (Herman and al., 1999) and Parameters mapping (Conversy, 2000). However the domain of sonification is still seldom used in Human Computer Interaction/Interface.

In literature there have been many projects that have tried to match images and sound. Distinguish among these projects is VoiceInteractive (Peter, 1996), which allows transferring grayscale images to sound thanks to the Fourier transformations. However, the sound produced by this mapping consists of a wide range of non-significant frequency variations that have no meaning to human

Please use the following format when citing this chapter:

Chemseddine, M. and Noirhomme-Fraiture, M., 2008, in IFIP International Federation for Information Processing, Volume 272; *Human-Computer Interaction Symposium*; Peter Forbrig, Fabio Paternò, Annelise Mark Pejtersen; (Boston: Springer), pp. 195–200.

perception. In another project, Weather sonification (Flowers and al., 2001), the correlation is made between different types of data, such as air pressure and temperature, with musical instruments. Here, the pitch is proportional to the value of data. Using an orchestration of data, Charlie Cullen and Eugene Coyle (2004) have proposed to sonify a database of employers using this mapping between data and musical melodies. For example, they propose to match the attributes of musical instruments or rhythmic and pitches to a tuple (database record). In an application domain close to our topic, Maria Barra (2000) presented her thesis in a project named Personal WebMelody(Taria and al., 2001) , which is a customizable system of sonification to monitor web servers that generates music through external sources (audio CD, MP3, and so on.). In this project the musical melodies are predefined in advance and are configurable by a webmaster.

Several studies have been conducted to try to find the best techniques to convey information. However, the problem of finding an appropriate correlation between data and sound is far from being solved. For some applications, the desire is to create realistic matches in the hope that they will be immediately "bindable", but also understandable.

The main purpose of this research is to design a sonification system that produces digital music from data in a meaningful and harmonic way. We intend to develop a prototype that will sonify networks traffic using music.

2. Relation between music and emotion

Music has an important role in attracting attention, transporting implicit and explicit messages, generating emotions and helping one retain information. These ties between music and emotion are part of the human psyche, primordial in nature. The psychologist Wilhelm Wundt (Population Research Laboratory, 2006) has described three dimensions of human emotion: valence, arousal, and potency. These dimensions are a subject of study in various fields of research in psychology, sociology and neurology.

The "Circumplex" by Russell (1989) (Schubert, 1999) is a classification of a human emotion within a two-dimensional model. It is limited to the two axes valence and arousal. The presentation of "Circumplex" suggests that the emotions happy and unhappy are opposed with respect to the axes of valence. The second dimension, arousal, measures the human activeness or passiveness in emotion.

In literature we see the work of Emery Schubert (1999), who summarized several scholarly studies in the field of musical emotion during the 19th and 20th centuries, proposing a model of the relationship between music elements and emotion. This two-dimensional emotional space based on the "circumplex" model and divided into 4 Quadrants, indicating a distribution into 8 clusters of human emotion (Figure 1)

Each cluster has its own musical element properties. In this work we limit ourselves to the following elements: articulation, vibrato, rhythm, meter, pitch, pitch range, mode, pitch combination, harmony, key and tempo.

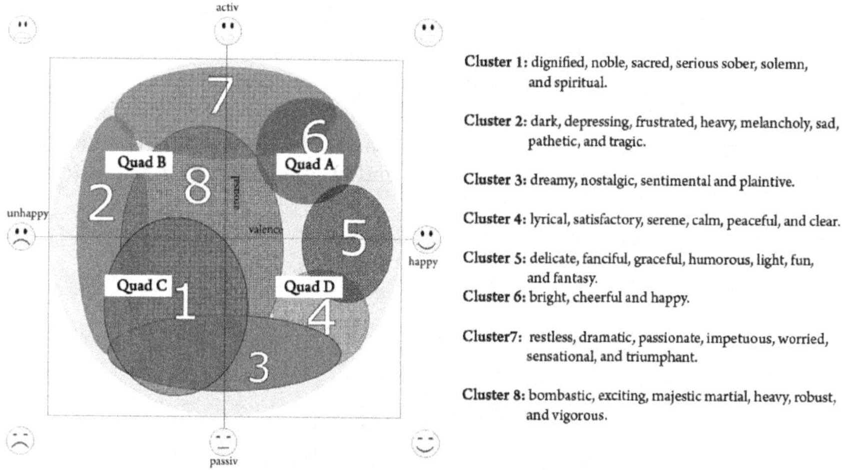
activ
7
6
Quad B
Quad A
8
arousal
2
unhappy
valence
5
happy
Quad C
1
Quad D
4
3
passiv
Cluster 1: dignified, noble, sacred, serious sober, solemn, and spiritual.
Cluster 2: dark, depressing, frustrated, heavy, melancholy, sad, pathetic, and tragic.
Cluster 3: dreamy, nostalgic, sentimental and plaintive.
Cluster 4: lyrical, satisfactory, serene, calm, peaceful, and clear.
Cluster 5: delicate, fanciful, graceful, humorous, light, fun, and fantasy.
Cluster 6: bright, cheerful and happy.
Cluster7: restless, dramatic, passionate, impetuous, worried, sensational, and triumphant.
Cluster 8: bombastic, exciting, majestic martial, heavy, robust, and vigorous.

Figure 1 two-dimensional emotional space source: E. Shubert (1999).

3. Data and emotion

Our next step is to match data with emotion, and to do that we must first choose an application. We have decided to test our method with network traffic control and supervision. The main goal of this application is to implement a real time system of sonification, which gives the administrator information about network traffic through mobile phones, PDA and so on. The mapping of data created by the Algorithmic Composition technique (Alpern, 1995) transforms data into digital music, based on emotional music elements. We have started state of the art research within the fields of sonification and musical emotion, with the goal of creating an interlinking model.

3.1 *Data and emotion groups*

This study has been divided into two parts. The first part aims to find a relation between music elements (which are mainly melodic, harmonic and rhythmic) and information. We started by interviewing 6 network administrators, in order to classify different messages displayed by network monitoring tools. We distinguished two groups of messages, the first is called "*Alert*" and the second is called "*Non-Alert*". The "*Alert*" group is composed of *Error, Warning, Critical,* and *Emergency* messages. The "*Non-Alert*" is composed of *Information, Notification* and *Debug*. Administrators affirmed that their emotional state towards a problem in network depends primarily on three factors:

- The impact of the problem on the user.
- The number of users infected by the problem.
- Their experience in dealing with such problem or similar ones.

The result of the interviews allowed us to have an idea about each message group's placement within the two-dimensional emotional space (Figure 2), finding that the representation of the group "*Alert*" occupies both quadrants B and C, and the group "*Non-Alert*" occupies quadrants A and D. According to that, clusters are distributed with respect to groups where:

- Group "*Alert*" occupies the clusters 1, 2, 3, 7, 8
- Group "*Non-Alert*" occupies the clusters 1, 3, 4, 5, 6, 7

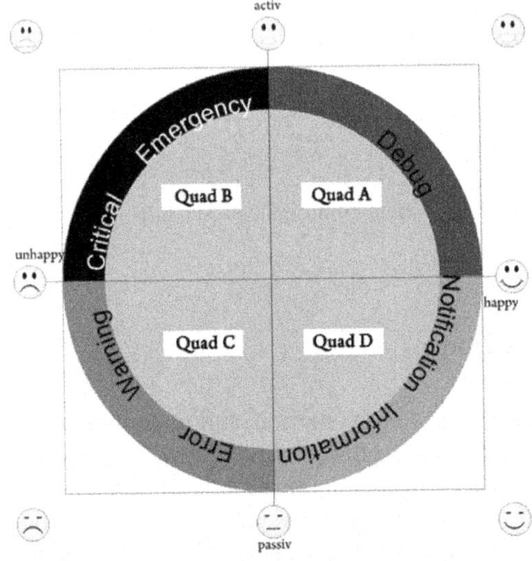

Figure 2 Classification of network message in two-dimensional emotional space.

A more precise placement will be concluded from a prepared survey planned to take place in the near future. Until then, with the help of this distribution along this two-dimensional emotional space and based on music elements of clusters (see section 2), we are still able to approach the different musical properties linked to each group.

Table 1. Example of music elements

message	Articulation	meter	pitch	Mode	Combination
Emergency Critical	Legato	triple	High / Low	minor	(1-2-3-4-5-6-5) / (5-6-5) / (5-1-2-3) / (1-2-3-2-1) / (5-3-2-1) / (1-2-3-4-5) / (1-2-3-2)
Warning	Legato	double	Low	minor	(5-4-3-2-1) / (8-7-6-5) / (descending chromatic) / (1-2-3-

Error					2-1) / (5-3-2-1) / (1-2-3-4-5) / (1-2-3-2)
Information	staccato	double	High / Low	major	(5-1-2-3) / (1-2-3-4-5) / (5-6-5) / (8-7-6-5) / (1-2-3-4-5-6-5) / (5-4-3-2-1)

3.2 *Online survey*

The second part consists of a multimodal (sound, graphic…) online survey via a website. Our goal here is to check our method, verify musical element in Table 2, and have a more precise classification of message subgroups. The human music composition used in the survey is based on music emotion element rules. Figure 3 represents a sample of musical partition of *Error, Critical* and *Information* messages (Chemseddine and Noirhomme, 2007). We have completed planning the experiment survey and selected 60 participants who are network administrators.

4. Hypothesis of our study

Our study is essentially based on "Western Music" and we have limited music elements in articulation; namely vibrato, rhythm, meter, pitch, pitch range, mode, pitch combination, key and tempo. Musicologists such as Byrd and Crawford (2001) (Kostek, 2005) argue that the most informative attributes in music are the melody and rhythm, allocating approximately 50% for melody and 40% for rhythm. The remaining 10% consists of harmony, dynamics, and articulation. Also, we have chosen to arrange music with a piano timber.

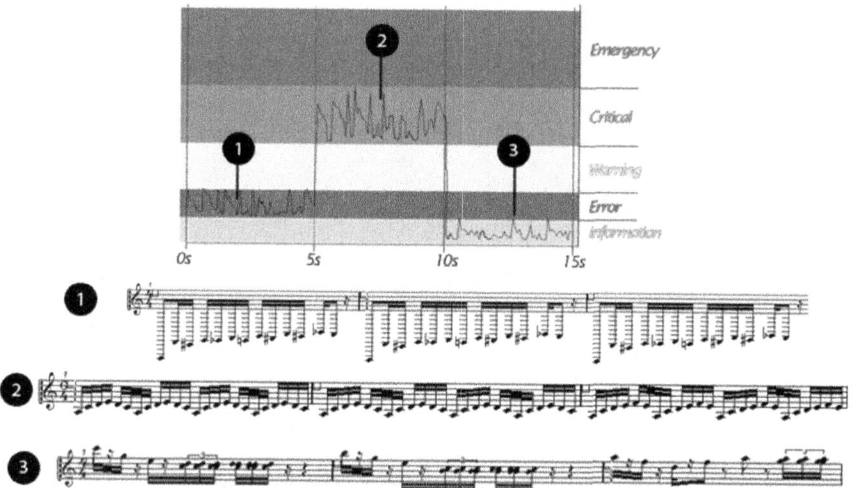

Figure 3 human music creations mapping graphic data

5. Conclusion and perspective

The next part of our work aims to develop algorithms for music composition and to test our application with users. We believe that our system can be used in various fields such as in the industrial production in processes control, in the sonification of content, in video games, in monitoring different networks (roads, computer, telephony etc.) and in musicotherapy.

References:

Alpern, A..: Techniques for Algorithmic composition of Music, Hampshire College (1995).

André, S., Peter, C., and Edwards, D.N.: An Evaluation of Earcons for Use in Auditory Human-Computer Interfaces, University of York (1993).

Conversy, S. : Conception d'icônes auditives paramétrées pour les interfaces homme-machine, pp. 43-46 (2000) .

Coyle, E., Cullen, C.: Orchestration within the Sonification of basic Data, Institute of Technology, Belfast (2004).

Chemseddine, M., Noirhomme, M.: Sound samples: http://netmelody.media19.be/samples.html (2007).

Flowers, J.H., Whitwer, L.E., Grafel, D.C. and Kotan, C.A.: Sonification of daily weather records: issues of Perception, attention and memory in design choices, University of Nebraska (2001).

Hermann, T., Ritter, H.: Listen to your Data: Model-Based Sonification for Data Analysis, University of Bielefeld (1999).

Hermann, T.: http://www.sonification.de/ (2006).

Kostek, B.: Perception-Based Data Processing in Acoustics, vol.3, p.30, Springer Heidelberg (2005).

Peter, B.L.: The Voice, http://www.seeingwithsound.com/javoice.htm (1996).

Population Research Laboratory: Valence and Arousal as Fundamental Dimensions of Emotional Experience (2006).

Schubert, E.: Measurement and time series analysis of emotion in music, Vol.1 (1999).

Tania, M.B., Antonio, C. and Matlock, T.: Personal WebMelody: Customized Sonification of Web Servers (2001).

Collaborative Knowledge Building for Decision-Support System Development

Helena Lindgren

Department of Computing Science, Umeå University, SE-901 87 Umeå, Sweden, helena@cs.umu.se

Abstract: The clinical domain of cognitive diseases and dementia is recognized by its highly complex knowledge domain, requiring expertise and experience in handling situations with a variety of symptoms and diseases, distributed over different levels in organizations and different professions. In this paper a pilot study is presented where eight experienced physicians in Sweden and Japan used an early prototype of the decision-support system DMSS (Dementia Management and Support System) in one to five well-known patient cases each. The prototype functioned as a mediator of a reflective conceptual artifact, i.e., the current understanding of the activity in focus in each patient case. The aim was to develop a common understanding of the clinical domain knowledge, differences in local process knowledge, needs for support and interactivity, by using the prototype as mediator. The physicians were observed using the system and interviewed individually and in groups. Results include adjustments of knowledge sources, terminology and design of user interface, interaction and knowledge base.

Keywords: Work analysis, clinical decision-support systems, knowledge-based systems, knowledge acquisition, participatory design, dementia care

1. Introduction

The clinical decision-support system (CDSS) DMSS (Dementia management Support System) is being developed for assisting medical personnel in the investigation of suspected cases of dementia (Lindgren, 2007). The main purpose of the system is to function as an extension of the individual actor's cognitive ability and as a common ground for collaborative and distributed teamwork (Singh, 2006). The system is designed to support higher-level cognitive functions such as reasoning, decision-making and learning in the sense of (Kaptelinin, 2005; Vygotsky, 1978). However, a critical functionality is also to collect patient data with high quality to be used to further develop the international evidence-based knowledge in the domain and the knowledge in the system. In this work it is of

Please use the following format when citing this chapter:

Lindgren, H., 2008, in IFIP International Federation for Information Processing, Volume 272; *Human-Computer Interaction Symposium*; Peter Forbrig, Fabio Paternò, Annelise Mark Pejtersen; (Boston: Springer), pp. 201–206.

outmost importance that the concepts and qualifiers used in the system bear the same meaning regardless country or local practice routines.

A common conception of how the investigation process is done and of the knowledge used in the process is needed in order to create a system, which fulfils the abovementioned purposes. A vital component is also a common view of the differences in work routines, treatment of concepts rooted in international evidence-based medicine, but also locally developed concepts and treatment protocols. In the development of this reflective conceptual artifact in the sense of (Singh, 2006), the inclusion of the prototype system and well-known patient cases are essential for serving as mediating tools for discussions. The method is qualitative and formative, thus suitable for work environments recognized by rapid development and change (Engeström, 1987; Häkkinen & Korpela, 2007; Korpela et al, 2001).

The results presented in this paper are used as a bench-mark for further development of the knowledge structures to be implemented in the system, for the design of interactive clinical reasoning, and as a base-line for evaluation studies in clinical practice.

2. Methods and Material

The study design was qualitative, with the aim of capturing as many aspects as possible in interviews and in sessions where physicians were observed using the system with patient cases. The physicians were asked to prepare up to five patient cases with suspected or established cognitive disease with information about the patient available on paper before the sessions. The patient cases formed use scenarios. The prototype system was introduced ("hands-on") at an open session. In the initial session one patient case was analyzed while aspects of the patient case, related to domain knowledge, local routines and the system were discussed. After that, individual sessions were held with physicians using the system with their patient cases in focus. A few of the physicians had used the system for a short period on in ongoing investigations of patients, or had tried the system with completed cases. The sessions were recorded on video and analyzed. The observers took an active role in the sessions and interviewed the physician during the use of the system. In some sessions additional physicians were participating in discussions during the sessions. Both Swedish and Japanese physicians participated in all group sessions.

All physicians participating in the study had profound experience in diagnosing and treating patients with dementia diseases and several have contributed with research in the domain. In their current work environments they were focusing different problem areas, therefore, their 24 patient cases represent both typical patients as well as extremely rare cases of dementia. The proportion was six cases of four different rare diagnoses for which support was not integrated in the system, eleven typical cases and seven atypical cases in the sense of (Lindgren, 2007).

3. Design of Interactive Clinical Reasoning

The DMSS system is designed to provide interactive support throughout the diagnostic process, giving reminders of what necessary evidence is missing, giving alerts when particular data requires alternative trails of investigation, giving suggestions of diagnosis, etc. This type of interaction behavior is also denoted mixed-initiative interaction in literature (Cortellessa & Cesta, 2006). All physicians in Japan systematically entered all information available about a patient, before using the analysis function. The way of interacting with the system by letting the system guide the gathering of only the necessary information was demonstrated to six of the physicians, who tested the method in one patient case each. Within the limited time to use the system in the study, the former systematic way of using the system seemed to be more natural, while the second was perceived to be a way to make the interaction faster.

The level of granularity of features and concepts in the system was discussed. The level of detail in laboratory findings and other/earlier diseases was found too coarse, since some of the physicians would like to enter specifics about some diseases other than cognitive diseases. Furthermore, a need for means to value the importance of different results from radiology examinations differently was also expressed, which would be beneficial especially in the presence of ambiguous evidence. This would correspond to the way two of the clinicians handled ambiguities in their patient cases in this study.

In the atypical cases the system presented the number of core features supporting different diagnoses possibly manifested in a patient case based on a set of certain clinical guidelines, without suggesting one particular diagnosis. This response of the system leaded the physician to reconsider his clinical diagnosis in a few cases. A need to be able to re-consider features and diagnoses in a patient case with support from the system was expressed. This need became obvious in the five cases treated by a Swedish physician, who recorded a preliminary diagnosis in the electronic patient record at the first encounter with the patient, despite the lack of evidence necessary for establishing a final diagnosis. Consequently, a critical feature of DMSS is its ability to support hypothesis generation during the investigation process.

The investigation of dementia diseases and the management of the progressive disease contain mainly i) diagnosis; ii) assessing the level of progress and severity in an individual patient at different points, including behavioral and psychological symptoms in dementia; iii) interventions; and vi) determining the level of aid the patient need in relation to the care provided by Japanese or Swedish health care system. Focus in the sessions was on diagnosis due to the patient cases and the limited scope of the prototype system. Four of the physicians who receive referred patients for diagnosis expressed that they use, or would like to use the system for verifying their clinical diagnoses, primarily in the difficult cases, and as a checklist for clinical investigations. A few of these also perceived the content of the system to be too rich for the primary care and too time consuming to enter all information that is needed. Other perceived the system to be a potential means to decrease the number of typical cases that is referred to experts, who should spend their knowledge on difficult cases. In these cases the content of the system needs to be to an extent rich in order to be a tool for education of primary care physicians not accustomed to investigate dementia.

It is more important to aid the assessment of severity and how to care for the patient than to aid diagnosis, in the perspective of the primary care physician who refers a majority of his patients for diagnosis. A common need expressed by all physicians is tools for assessing behavioral and psychological symptoms and for support on how to handle the care for persons with such symptoms.

Two of the physicians had access to the system during a period of one month before the evaluation sessions were held. The physicians used the content of the system printed on paper as a checklist in the encounters with new patients, and verified diagnoses by using the system after the patient data was collected. As such, they found the system valuable, especially in difficult cases.

To summarize, the usefulness and purpose of the system differs, mainly depending on whether the local routines at a certain clinic include diagnosis and has access to advanced equipment or not. Therefore, the system should have mechanisms to distinguish and support also activities other than diagnosis, in order to provide the level of support needed in the local practice.

4. Expert's Assessment of Diagnosis

The proportion of rare cases was high in the limited amount of patient cases in this pilot study, due to the range of expertise participating in the study (seven cases of 24 in total). Statistically, these dementia types represent no more than a few percent of all dementia cases. These cases were also perceived as difficult cases and some of them had been misdiagnosed during the progress of the disease. The limited version of DMSS used in the study was designed for supporting primarily typical, unambiguous dementia cases of the most common types for the purpose of serving the inexperienced primary care physician. Hence, the cases with rare diagnoses were included for investigating the behavior of the system in cases not supported by the system. In only one of the cases the system produced a satisfactory result when reporting that the evidence was incomplete and further investigation was needed. In two of the cases the system acknowledged the complicated situation and presented support for possible diagnoses. Obviously, there is a need to integrate support for also detecting these rare cases.

The system's suggestions of diagnoses correlated with the clinical diagnoses in all typical cases, except for one case due to the different clinical guidelines that was used for diagnosis on a routine basis. When the system applied the same guideline as the clinician, the system reached the same results.

In the seven *atypical* cases the system did not suggest one single diagnosis since the evidence was ambiguous in these cases in the perspective of a certain set of guidelines (Lindgren, 2007). Instead, the system presented a summary of critical features for each diagnosis for which clinical guidelines were implemented in the system. The dominant situation in these cases was that the evidence supported two diagnoses. Three of these were patients currently living in group homes and had the clinical diagnosis Alzheimer's disease. These cases were cases in which the disease had been proceeding further than in the other cases, a conclusion which was supported by the score they had on the Hasegawa scale. In these cases more symptoms of a various kind had been developed, which in distribution resembles the alternative diagnosis. Whether this is an indication of that the system is more suitable for diagnosis early in the development of the disease, or that the system actually identifies complicated cases that have been misdiagnosed in earlier stages, needs to be further investigated. In one of these cases, the physician explicitly was going to use the response from the system to reconsider the clinical diagnosis.

5. International Knowledge vs. Local Praxis

The results showed no differences in how the main diagnostic procedure is processed by individual physicians when compared to international clinical guidelines and between the two countries (Lindgren, 2007). Basically, the same clinical guidelines are used in both countries. However, differences were expressed in what way the guidelines were used. For instance, one guideline was used in Japan on a regular basis, while in Sweden (and in DMSS) this guideline is used primarily in research and in difficult cases. This caused the different diagnostic outcomes in one patient case in the study. There is also a difference also in what screening tools are being used for collecting basic data in routine practice.

Furthermore, there are differences in how certain features and concepts are used in clinical practice in the different countries. These are either due to language structures, locally developed medical concepts or different usage of concepts rooted in international evidence-based studies. In order to make the data useful for research purposes, such distinctions need to be clarified and implemented in a way that the inferences become valid and the meaning of data is the same, regardless if it is collected in Japan or in Sweden.

6. Conclusions

A participatory assessment of what support is needed in dementia care is described, using a prototype system and well-known patient cases as means in sessions with experienced physicians of different expertise in Sweden and Japan. The system, used for capturing the patient case, functioned as a mediator of a reflective conceptual artifact, i.e., the understanding of the activity in focus. The aim for the case study was to assess differences in reasoning and work processes, and in what resources are used when investigating dementia in different clinics in Sweden and Japan. Furthermore, the role of DMSS in its current version was investigated, as well as its potential role in a developed version. Results include different preferences in what clinical guidelines to base diagnosis upon, in what way the system is used, and terminology issues to be solved. The physicians had similar view on the need of a decision support system in their daily work and in what form this support should be provided. The expected benefits were related to the amount of experience in an individual physician and to what extent the system supports the management of difficult cases.

References

Cortellessa, G., and Cesta, A.: Evaluating Mixed-Initiative Systems: An Experimental Approach. In ICAPS-06. Proceedings of the Sixteenth International Conference on Automated Planning and Scheduling pp. 172-181. Menlo Park, CA: AAAI Press (2006).

Engeström, Y.: Learning by expanding: An activity-theoretical approach to developmental research. Helsinki, Orienta-Konsultit (1987).

Häkkinen, H., and Korpela, M.: A participatory assessment of IS integration needs in maternity clinics using activity theory. Int J Med Inf 76:843-849 (2007).

Kaptelinin, V.: Computer-mediated activity: functional organs in social and developmental contexts. In B. Nardi (Ed.), Context and consciousness: activity theory and human-computer interaction. MIT, Cambridge, USA pp. 45-68 (1995).

Korpela, M., Soriyan, H.A., and Olufokunbi, K.C.: Activity analysis as a method for information systems development. Scandinavian Journal of Information Systems 12(1-2):191–210 (2001).

Lindgren, H.: Decision Support in Dementia Care – Developing Systems for Interactive Reasoning. (2007) http://urn.kb.se/resolve?urn=urn:nbn:se:umu:diva-1138. Accessed 4 Feb 2008.

Patel, V.L., Arocha, J.F., Diermeier, M., How, J., and Mottur-Pilson, C.: Cognitive psychological studies of representation and use of clinical practice guidelines. Int J Med Inf 63:147-167 (2001).

Singh, G.: Investigating the support of reflective activities by collaborative technologies: an activity theory based research model. In: Proc. TT211C2006 pp. 49-58 (2006).

Vygotsky, L.: Mind in society: The development of higher psychological processes. Cambridge, Harvard University Press (1978).

Multitouch Sensing for Collaborative Interactive Walls

Alessandro Soro[1], **Gavino Paddeu**[2], **and Mirko Luca Lobina**[3]

[1] CRS4, Italy, asoro@crs4.it
[2] CRS4, Italy, gavino@crs4.it
[3] CRS4, Italy, mlobina@crs4.it

Abstract: In this paper we explain the design of t-Frame, a hardware/software architecture that allows the implementation of multiuser interactive wall. t-Frame brings multi-touch sensing to a generic display by means of low cost digital video cameras. The design of t-Frame is illustrated in detail, together with a prototype installation. We show how t-Frame differs from other approaches and discuss our findings, together with a plan of future research and improvements.

Keywords: pervasive, multi-user, multi-touch, interactive wall.

1. Introduction and Motivation

Interactive walls are a special kind of computer applications that deliver a highly impressive, shared view of information, and are suited to many exciting applications, ranging from workgroup collaboration to pervasive computing and entertainment. t-Frame is a low-cost hardware/software architecture that enables multi-touch interaction on a generic display. Specifically, t-Frame is intended to be used in large size, multi-user interactive walls. With respect to other approaches t-Frame can be installed as a pointing device on any flat surface regardless of size, shape and display technology.

t-Frame allows researchers in the field of human-computer interaction to set up with minimal effort an environment for experimenting in the field of computer supported collaborative work and tangible user interfaces: the goals of the project can be summarized as follows:

1. provide multi-user and multi-touch interaction to any display, regardless of the specific technology of the display itself;
2. minimize both the cost of installation and maintenance, using standard hardware and simplify the installation and calibration procedures;

Please use the following format when citing this chapter:

Soro, A., Paddeu, G. and Lobina, M.L., 2008, in IFIP International Federation for Information Processing, Volume 272; *Human-Computer Interaction Symposium*; Peter Forbrig, Fabio Paternò, Annelise Mark Pejtersen; (Boston: Springer), pp. 207–212.

3. be applicable to very wide installations: modern hardware allows the creation of interactive walls several meters long using cluster systems or multi-head graphic adapters.

By contrast, the most adopted technique to implement a multitouch displays is, at the moment, based on FTIR (Han, 2005), and requires the adoption of an expensive high resolution IR camera, ambient IR screening and is in practice bounded to rear projected screens.

t-Frame requires less space than other technologies, can be easily transported because it does not require a single-piece touch surface (the prototype implementation described in this paper consists of a display tile, although any other display solution can be exploited) and, most important, can adapt to the size of the display seamlessly, without sensibly affecting the performances.

2. Related Work

Pioneer work on multi-touch sensing devices can be tracked back to the mid eighties, see for example (Lee, 1985)(Krueger, 1985)(Kasday, 1984). An overview of the evolution of multi-touch technologies is maintained in (Buxton, 2008). Given that multi-user interaction is a straightforward extension of multi-touch sensing, the obvious playground in this field consists in displays capable of accommodating a number of users, such as tabletop and wall-size displays. (Dietz, 2002) and (Wilson, 2005) are examples of the former, (Wilson, 2004) and (Dempski, 2005) of the latter. Several techniques have been exploited to implement multi-touch sensing devices: (Dietz, 2002) consists of an array of antennas whose signals get transmitted, through the body of the user, to a receiver that elaborates touch events. Among optical techniques (Wilson, 2004) exploits stereo cameras to compute hands position, but the cameras are located behind the semi-transparent screen, thus the system is bounded to front/rear projected display. The same holds in (Han, 2005), which relies on an infrared camera that captures the light that escapes the display surface when finger contact occurs. In (Oka, 2002) the optical sensor is located above the display surface, and thus the hands of the user(s) stay between the camera and the screen. Although feasible for the interactive desk described, this approach would be not practical for interactive walls, since the body of the user would in general cover the hands from the viewpoint of the camera. The system described in (Denlinger, 1988) exploits an approach similar to t-Frame: the cameras are arranged around the screen and the position of the fingers is determined through triangulation, but the cameras are located on the corners of the screen: in t-Frame the particular arrangement of the cameras limit the complexity of the algorithms involved in finger triangulation and allows the system to scale in size almost indefinitely.

3. t-frame Design

A t-Frame installation consists of a set of cameras arranged on the plane of the display. In a typical installation of an interactive wall, cameras are aligned on top of the screen, facing downwards, as shown in Figure 1a. However, cameras are not bounded to a fixed position or orientation, and can be arranged anyhow, as long as they lie on the same plane of the screen and the exact position and orientation of each camera is known with respect to screen coordinates. In order to simplify the setup of the installation t-Frame provides a calibration facility that computes the exact position of each camera, this operation involves three steps: i) every camera takes a snapshot of its field of view, no touch must occur during this process, and saves it as a known background; ii) the user is requested to point an horizon in the background image of each camera: the horizon must be specified as close as possible to the surface of the screen; iii) the user is requested to point with her finger three given points on the screen: the position of the cameras is triangulated exploiting the images captured at each touch. Once the installation has been correctly calibrated each camera enters a continuous loop in which still images are captured at the maximum possible frame rate. Periodically the images are compared to gather touch events that are then pushed in an event queue that applications can poll and consume. In the following the two most critical steps of t-frame are presented.

3.1 Finger Triangulation

The frames captured by each camera are elaborated to spot touch events. A frame is compared against the known background along the line of the horizon: when a significant difference is found, the algorithm assumes that the background is covered by a finger touching the screen (see Figure 1b) and measures its position. The position of the finger is reported as the angle under which the finger is seen with respect to the center of the field of view. To do this the exact aperture of the field of view must be known, since we only can take measures in terms of relative positions, i.e. counting pixels in the image. Figure 1c shows how the exact position of a finger is computed in screen coordinates: when a finger touches the surface of the screen it is seen by every camera whose field of view covers that position. Additionally every camera sees the finger under a given angle. Then, computing the position of the finger is as easy as calculating the intersection of two lines passing through the position of any given camera (that is known from calibration), and intersects the axes of the screen under the given angle, which is a matter of bare trigonometry. With a single touch and only two cameras this approach doesn't differ from stereo vision, but by exploiting several cameras t-Frame can easily recognize an arbitrary number of touches.

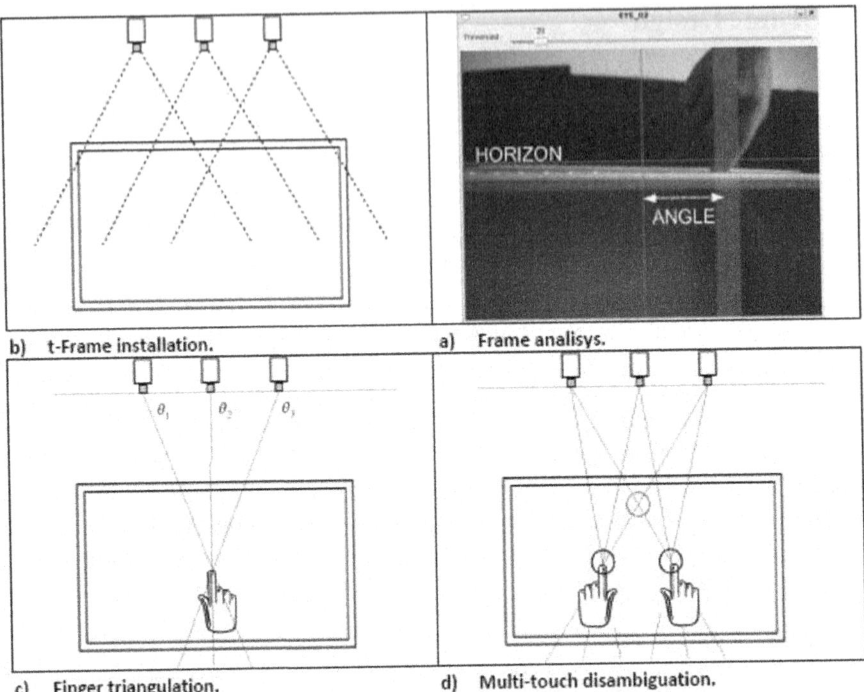

Figure 1 t-frame design.

3.2 *Multi-touch disambiguation*

Consider the situation schematized in Figure 1d. The user touches the screen with two fingers: each camera sees its background covered by two distinct obstacles and computes two angles. We can therefore trace two lines for each camera; the intersections of such lines determine the exact position of the fingers, but we need a strategy to exclude false spots. To this end the algorithm of multi-touch disambiguation compares every intersection (that is a candidate finger) against any camera that has that specific point within its field of view, and checks if the candidate finger is compatible with the image seen. In the case of Figure 1d the candidate finger highlighted with a grey circle is the result of the analysis of the frames taken by the leftmost and rightmost cameras. In order for this to be accepted as a finger-touch, it must be confirmed by the frame taken by the central camera. The comparison against the central camera shows that this is a fake finger, as expected, and that only the two spots highlighted in black circles represent true finger-touches. The number of cameras needed to disambiguate a given configuration depends on many factors. In general with N cameras we can always disambiguate N – 1 finger-touches, but in practice fewer cameras are enough to monitor many finger-touches. In fact, in many cases, false candidates fall (largely) outside of the display area and can be excluded. Additionally finger-touches tend to follow predictable paths, and heuristic techniques can be exploited to distinguish false candidates. It can be shown that the complexity of the algorithms is linear with respect to the number of cameras installed, and depends primarily on the number of fingers that are seen at once by a given set of adjacent cameras. A formal proof of this result is anyway outside the scope of this paper.

4. Prototype Implementation

The t-Frame design and algorithms have been tested in a prototype installation, consisting of a 60" wide display tile, driven by a high-end graphical workstation. The optical apparatus consisted of 9 VGA cameras (25 frames/sec). The system performs very well in terms of efficiency, since the algorithms involved in the image analysis don't comport a significant overload. In addition, the exact computation of fingers' position only involves some basic operations. Our experiments show that multi-camera triangulation based on background subtraction is a suitable technique for implementing a multi-user & multi-touch interactive wall. The event models that are common in computer systems, based on clicks, drags and so on, have been designed for single pointer operations. Although such a model can be extended to support multi pointer interaction (such as in MPX http://wearables.unisa.edu.au/mpx/) this cannot be other than a first step towards a new, commonly agreed, event model specifically designed for multi-touch widgets.

5. Conclusions and Further Work

We have shown the design and a prototype implementation of t-Frame, a computer architecture that deploys multi-touch sensing capabilities to any display surface. This novel framework presents several encouraging aspects: i) it is a completely new approach for implementing a multi-touch strategy; ii) it is fully scalable from small touchable surfaces to large interactive walls; iii) it can be applied also to no-display surfaces, as geo-charts or posters; iv) it is basically a solution of compact dimensions, easily portable and based on cheap hardware. The next step of the t-Frame project is the clarification of several shortcomings we have observed during experiments. First of all the precision of the pointing system strongly depends on the resolution of the installed cameras. At the same time also a certain temporal data-ambiguity is observable due to the frame rate of the cameras. Finally, a new paradigm should be introduced for the event management. In fact, the typical model of events introduced with the mouse is not suitable for a multi-touch framework. Apart from the clarification of these problems, we are particularly interested in experimenting in the near future with no-flat screens, such as panoramic, or even cylindrical, displays, either front and back projected.

Our intention is to deploy t-Frame as collaboration platform for complex information retrieval and manipulation, but besides our initial scenarios, many other applications are emerging that could take advantage from the technology described in this paper, in particular, outdoor installation of public interactive walls, to use as high-tech exhibition spaces, and entertainment platforms.

Acknowledgment:
The t-Frame project has been partially funded by the Italian Ministry of University and Research (MIUR) as part of the project DART(Angioni, 2007) (Distributed Agent-Based Retrieval Toolkit), contract grant number 11582.

References

Angioni, M., Demontis, R., Deriu, M., De Vita, E., Lai, C., Marcialis, I., Pintus, A., Piras, A., Soro, A., Tuveri F.: User Oriented Information Retrieval in a Collaborative and Context Aware Search Engine. WSEAS Transactions on Computer Research, ISSN: 1991-8755, vol. 2(1), pp. 79-86, (2007).

Buxton , W.: Multi-Touch Systems that I Have Known and Loved. http://www.billbuxton.com/multitouchOverview.html. Accessed 15 Apr 2008.

Denlinger, B. M. Ambient-light Responsive Touch Screen Data Input Method and System. U.S. Patent 4782328. (1988)

Dempski, K., Harvey, B.: Supporting Collaborative Touch interaction with High Resolution Wall Displays. In 2nd Workshop on Multi-User and Ubiquitous User Interfaces. Jan. 2005 at IUI'05. (2005)

Dietz, P. H., Darren, L. L.: Multi-User Touch Surface. U.S. Patent 6498590. (2002)

Han, J. Y.: Low-cost multi-touch sensing through frustrated total internal reflection. In Proceedings of the 18th Annual ACM Symposium on User interface Software and Technology (Seattle, WA, USA, October 23 - 26, 2005). UIST '05. ACM Press, New York, NY, 115-118. (2005)

Kasday, L.: Touch Position Sensitive Surface. U.S. Patent 4484179. (1984)

Krueger, M. W., Gionfriddo, T., and Hinrichsen, K.: VIDEOPLACE—an artificial reality. In Proceedings of the SIGCHI Conference on Human Factors in Computing Systems (San Francisco, California, United States). CHI '85. ACM Press, New York, NY, 35-40. (1985)

Lee, S., Buxton, W. & Smith, K.C.: A Multi-Touch Three Dimensional Touch-Sensitive Tablet. In Proceedings of the 1985 Conference on Human Factors in Computer Systems, CHI '85, San Francisco, April, 1985, 21-26 (1985)

Oka, K., Sato, Y., Koike, H.: Real-Time Tracking of Multiple Fingertips and Gesture Recognition for Augmented Desk Interface Systems. FGR 2002: 429-434. (2002)

Wilson, A. D.: TouchLight: an imaging touch screen and display for gesture-based interaction. In Proceedings of the 6th international Conference on Multimodal interfaces (State College, PA, USA, October 13 - 15, 2004). ICMI '04. ACM Press, New York, NY, 69-76. (2004)

Wilson, A. D.: PlayAnywhere: a compact interactive tabletop projection-vision system. In Proceedings of the 18th Annual ACM Symposium on User interface Software and Technology (Seattle, WA, USA, October 23 - 26, 2005). UIST '05. ACM Press, New York, NY, 83-92. (2005).

Visualization of Personalized Faceted Browsing

Michal Tvarožek and Mária Bieliková

Institute of Informatics and Software Engineering, Faculty of Informatics and Information technologies, Slovak University of Technology, Ilkovičova 3, 842 16 Bratislava, Slovakia, {tvarozek,bielik}@fiit.stuba.sk

Abstract: Current user needs and expectations increasingly shift focus away from simple information lookup towards more complex tasks collectively described as exploratory search. This requires the development of novel approaches and tools to search, navigation and processing of information, such as personalized faceted browsing. We describe a visualization approach for personalized faceted browsers suitable for exploratory search tasks in large information spaces, and apply it to the job offers, scientific publications and digital images domains.

Keywords: personalized faceted browsing, visualization, adaptation

1. Introduction

Contemporary applications and users deal daily with huge amounts of data on a daily basis, such as deep web (relational) databases and Semantic Web repositories, or miscellaneous web resources available via web search engines. While effective access to these resources via search and navigation is steadily improving, it can still be argued that the growing size and complexity of the available information space hampers overall user experience.

Moreover, growing user needs and expectations together with advances in information retrieval approaches shift the focus from simple lookup tasks to more complex tasks collectively described as *exploratory search* (Marchionini, 2006). These include *learning* (e.g., comprehension or comparison) and *investigation* (e.g., analysis, discovery, forecasting or transformation). This trend is further emphasized by the younger "Net generation" (i.e., people who grew up with the Web) as their needs, expectations, attitudes and information seeking behaviour are significantly different compared to the "pre-Web generation" (Oblinger, 2005).

In order to address issues of effective information access and processing, we propose information retrieval, human computer interaction, user interfaces, and initiatives such as the Semantic Web and the Adaptive Web as prime candidates for cross-fertilization of approaches for the creation of successful solutions. We

---

Please use the following format when citing this chapter:

Tvarožek, M. and Bieliková, M., 2008, in IFIP International Federation for Information Processing, Volume 272; *Human-Computer Interaction Symposium*; Peter Forbrig, Fabio Paternò, Annelise Mark Pejtersen; (Boston: Springer), pp. 213–218.

take advantage of semantic search in the Semantic Web environment (Guha, 2003) and provide users with an adaptive faceted browser interface as means for integrated search, navigation and exploration of the available information space.

2. Related Work

Exploratory search tasks involve a broader range of activities than traditional lookup tasks. Users thus need quick and easy access to more advanced methods, tools and features for information search, processing and understanding.

Advanced means for query construction and modification facilitate exploratory search, e.g., view-based search based on faceted browsers was proposed and successfully evaluated in various scenarios (Yee, 2003; Wilson, 2006).

Adaptive user interfaces dynamically adjust system behaviour to specific conditions, which might include device or environment properties, user characteristics or social relations. Typical applications include automatic user interface generation (Dakka, 2005; Oren, 2006) or personalization (Tvarožek, 2007), content adaptation or recommendation, e.g. in educational systems (Brusilovsky, 2004).

Authors in (Wilson, 2008) compare three major faceted browsers which allow users to formulate queries via navigation by successively selecting metadata terms in a set of available facets, and to interactively browse the corresponding search results. mSpace is a domain specific browser of RDF data, which provides users with a projection of high dimensional information spaces into a set of columns (filters), which can be manually added, rearranged or removed by users (Wilson, 2006). Flamenco stresses interface design and guides the user through the information seeking process from a high level overview through query refinement and results preview to the exploration of individual results (Yee, 2003). The BrowseRDF faceted browser supports facet generation from arbitrary RDF data and extends the expressiveness of faceted browsing by providing additional operators in addition to selection and intersection (Oren, 2006).

While in Flamenco the facets are static and predefined, users can manually adapt columns in mSpace to match their needs. BrowseRDF automatically identifies facets in source data based on several statistical measures, yet does not directly address issues of information overload or interface usability and adaptivity.

3. Visualization of Personalized Faceted Browsing

We proposed the extension of “classical” faceted browsers with personalization support based on user characteristics in (Tvarožek, 2007). Our goal was to improve navigation in large/complex information spaces by providing adaptive navigation guidance and orientation support aimed at the specific needs and requirements of individual users ultimately improving overall user experience.

We developed a personalized faceted browser – Factic, which allows users to navigate Semantic Web data collections represented in OWL repositories. Factic belongs to a larger platform aimed at personalization and presentation of semantically enriched data, acquired (semi)automatically from various web resources (Barla, 2007). Factic constructs the user interface based on a faceted classification extracted from the domain ontology and is thus effectively domain independent, provided that useful facets can be generated to populate the user interface.

To address exploratory search in the (Semantic) Web, our faceted browser was primarily designed for large information spaces containing millions of information artefacts with many available facets and restrictions imposing high requirements on the cognitive abilities of users. We assume high structural complexity – many different concepts with complex relations, and information space dynamics indicating frequent and/or unexpected changes in the data and the structure of the information space, also assuming missing or unknown data and metadata about items. We also address high user diversity – many different users with specific backgrounds, needs and levels of expertise.

The main focus of this paper lies in the visualization of personalization of facets and restrictions, which are the users' primary means of navigation in a faceted browser. Therefore, we employ existing successful faceted browser interface concepts such as the overall interface layout with facets on the left, the current query at the top and the search results in the centre (see Figure 1).

We use different kinds of facets based on the data types they operate on (e.g., date, numbers, enumerations, taxonomies) using these personalization steps:

1. *Select available facets* – evaluate the set of available facets and add or remove facets from the global facet pool to the individual user's facet pool, optionally constructing new facets from available metadata via dynamic facet generation.
2. *Determine facet types* – active facets are fully functional and their contents are visible, inactive facets are used for queries, yet their contents are hidden, disabled facets are not used for querying and their contents are hidden. However, inactive and disabled facets and their contents can be access on demand.
3. *Order facets and restrictions* – adapt the facet ordering based on descending relevance and usage statistics, orders the restrictions in facets either alphabetically, by relevance or by their size (number of instances they correspond to).
4. *Recommend facets and restrictions* – evaluate the relevance, expressiveness and "usefulness" of items, recommend the most suitable for further navigation.
5. *Annotate facets and restrictions* – add additional (domain specific) clues and information about individual items (e.g., the size of restrictions, explain the meaning of facets and restrictions, show examples of "what is behind them").

We distinguish individual facet types via background colour. Facets which were already used – a restriction was selected are shown in green, while blue facets are yet unused – without any selected restrictions. All restrictions in active facets are shown, while only the headers of inactive and disabled facets are visible. Two additional icons serve as buttons and indicators allowing users to activate/enable inactive and disabled facets respectively (see Figure 1, left).

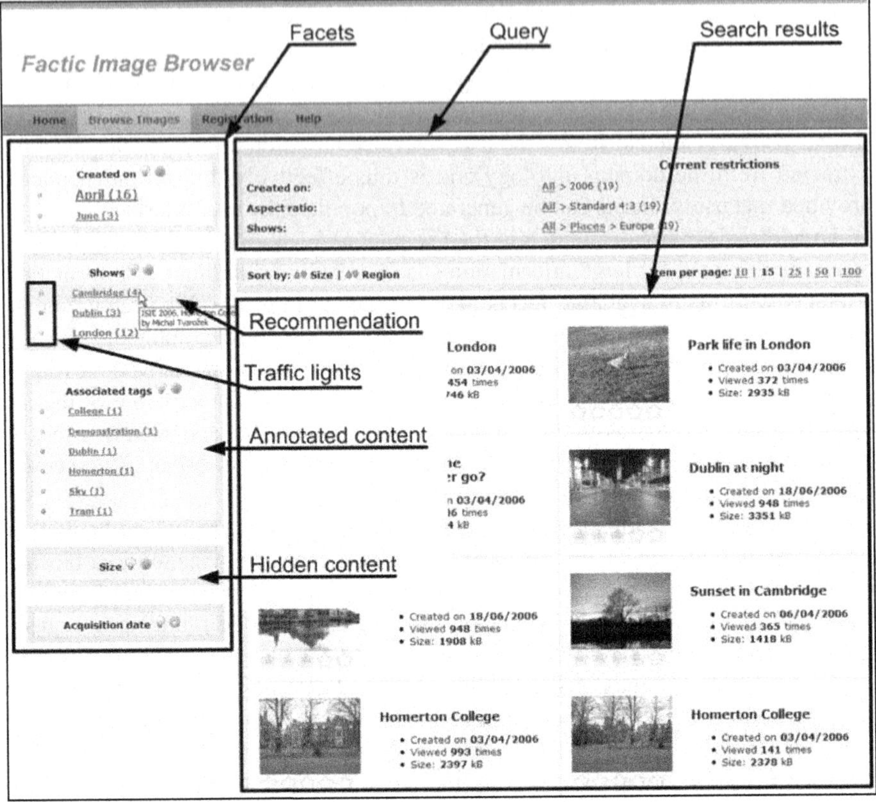

Figure 1 Example of our personalized faceted browser visualization in the digital image domain. Active/inactive facets are shown with different background colours; icons in facet headings indicate facet type and allow users to activate/enable facets on demand. The traffic light metaphor describes restriction relevance, font size indicates the relative number of instances that satisfy restrictions, and numbers indicate the absolute number of results. Restriction background colour denotes recommendations, while tooltips contain additional annotations (e.g., metadata descriptions or (personalized) content summarization).

While the visualization of restriction relevance via ordering was at first a tempting idea, our initial experiments revealed that it was not well accepted by users who were confused by the rather erratic ordering of items (e.g., due to weak relevance estimates) and inability to search them in a natural way.

Consequently, we decided to always sort restrictions alphabetically and display relevance and additional information via link annotation. We show *restriction relevance* using the traffic light metaphor, which has already been successfully used in several adaptive systems (Brusilovsky, 2004). Here, green corresponds to high relevance, yellow for medium relevance and red for low relevance (i.e., less suitable for further navigation). Similarly, in addition to the traditional number of search results corresponding to individual restrictions, we present restrictions with progressively different font sizes based on the relative number of instances that satisfy them. Thus, the most "rich" restrictions are quickly visible somewhat corresponding to sorting based on restriction size (see Figure 1, left).

To compensate for deep and/or complex faceted classifications and to provide shortcuts to popular items we recommend specific restrictions to users based on their estimated characteristics and social groups. These are presented at the top of the restriction list in each facet with green background colour. We can recommend both restrictions at the current level or deeper in the restriction hierarchy. In order to allow users to search the restriction list while having all recommended restrictions at the top, we duplicate recommended restrictions if they are at the current level of the facet hierarchy. We annotate facets and restrictions via tooltips with their descriptions retrieved as metadata from the used domain ontology, while also summarizing the contents of restrictions, e.g. by presenting some attributes of the corresponding search results that satisfy a given restriction (see Figure 1, left).

Lastly, we provide users with personalized visualizations with additional annotations, which provide different levels of abstraction via hierarchical cluster views, present fewer or more attributes of individual instances in adaptive views, and provide varying levels of navigation guidance and orientation (Tvarožek, 2008). Search results are normally unordered, or ordered via specialized external tools for single-/multi-criteria ordering (Gurský, 2005). Alternatively, context-based (keyword) search can be used to order/search the entire collection as in (Návrat, 2007).

Furthermore, we use horizontal incremental graph navigation in/visualization of concepts, instances and relations (Bieliková, 2007). E.g., the size, relative layout and colour of clusters provides information about instance counts, similarity, relevance (e.g., via a user's social network) and overall suitability (e.g., via user characteristics).

3. Conclusions

We proposed the visualization of an adaptive faceted browser interface suitable for the browsing of large open information spaces and developed the faceted browser Factic for search and navigation in Semantic Web repositories.

We performed experiments in three application domains – job offers (project NAZOU, nazou.fiit.stuba.sk), and scientific publications and digital images (project MAPEKUS, mapekus.fiit.stuba.sk). E.g., our publication ontology was populated with metadata acquired from ACM DL, SpringerLink and DBLP totalling about 985,000 publications (~140k from ACM DL, ~35k from SpringerLink and ~810k from DBLP) which were conceptualized into about 390 classes.

We perform (semi)automatic user interface generation based on a supplied domain ontology describing both the structure of an information space (i.e., metadata ~ facets) and data (i.e., individual information artefacts ~ search results). We address issues of large, open or inconsistent information spaces via adaptation employing usage statistics, user preferences and social navigation observations. Since both the visualization and the personalization approaches are effectively domain independent, our browser can be used in other application domains for which a domain description is available.

Acknowledgments This work was partially supported by the Slovak Research and Development Agency under the contract No. APVT-20-007104, the State programme of research and development under the contract No. 1025/04 and the Scientific Grant Agency of Slovak Republic, grant No. VG1/3102/06.

References

Barla, M., Bartalos, P., Bieliková, M., Filkorn, R., Tvarožek, M.: Adaptive portal framework for Semantic Web applications. In: Proc. of 2nd Int. Workshop on Adaptation and Evolution in Web Systems Engineering, ICWE 2007 Workshops, pp. 87-93 (2007).

Bieliková, M., Jemala, M.: Adaptive Incremental Browsing of Ontology Structure. In: Proc. of the 18th ACM Conf. on Hypertext and Hypermedia, UK, ACM Press, 143-144 (2007).

Brusilovsky, P.: Adaptive navigation support: From adaptive hypermedia to the adaptive Web and beyond. Psychology 2, 1:7-23 (2004).

Dakka, W., Ipeirotis, P.G., Wood, K.R.: Automatic Construction of Multifaceted Browsing Interfaces. In: Proc. of the 14th ACM Int. Conf. on Information and knowledge management, ACM Press, New York, NY, USA, 768-775 (2005).

Guha, R., McCool, R., Miller, E.: Semantic Search. In: Proc. of the 12th Int. Conf. on World Wide Web. ACM Press, New York, NY, USA, 700-709 (2003).

Gurský, P., Lencses, R., Vojtáš, P.: Algorithms for user dependent integration of ranked distributed information. In: Proc. of TED Conference on e-Government, 123-130 (2005).

Marchionini, G.: Exploratory search: from finding to understanding. Communications of the ACM 49, 4:41–46 (2006).

Návrat, P., Taraba, T.: Context Search. In: Y. Li et al.: Proc. of Int. Conf. on Web Intelligence and Intelligent Agent Technology (Workshops), IEEE CS, USA, 99-102 (2007).

Oblinger, D., Oblinger, J.: Is it age or IT: First steps toward understanding the Net generation. Educating the net generation. Educause, www.educause.edu/educatingthenetgen (2005).

Oren, E., Delbru, R., Decker, S.: Extending Faceted Navigation for RDF Data. In: Proc. of the 5th Int. Conf. on Semantic Web. LNCS 4273, Springer, Heidelberg, 559-572 (2006).

Tvarožek, M., Bieliková, M.: Personalized Faceted Navigation for Multimedia Collections. In: Proc. of SMAP 2007, 2nd Int. Workshop on Semantic Media Adaptation and Personalization. CS IEEE Press, 104-109 (2007).

Tvarožek, M., Bieliková, M.: Collaborative Multi-Paradigm Exploratory Search. In: Proc. of Web Science Workshop at Hypertext 2008, ACM Press, NY, USA, (2008), to appear.

Yee, K.P., Swearingen, K., Li, K., Hearst, M.: Faceted metadata for image search and browsing. In: Proc. of the SIGCHI Conf. on Human Factors in Computing Systems, ACM Press, New York, NY, USA, 401-408 (2003).

Wilson, M.L., schraefel, m.c.: mspace: What do numbers and totals mean in a flexible semantic browser. In: 3rd Int. Semantic Web User Interaction Workshop at ISWC2006 (2006).

Wilson, M.L., schraefel, m.c., White, R.W.: Evaluating Advanced Search Interfaces using Established Information-Seeking Models. In: Journal of the American Society for Information Science and Technology (JASIST), (2008), to appear.